Fueling our Future

A Dialogue about Technology, Ethics, Public Policy, and Remedial Action

A Circle of Discernment Report from Quaker Institute for the Future and Earthcare Working Group of Philadelphia Yearly Meeting of the Religious Society of Friends (Quakers)

Ed Dreby and Keith Helmuth, Coordinators
Judy Lumb, Editor

March 2009

—Quaker Institute for the Future Pamphlet 1—

Copyright © 2009 Quaker Institute for the Future

Published for Quaker Institute for the Future by *Producciones de la Hamaca*, Caye Caulker, Belize <judylumb.com>

ISBN: 978-976-8142-20-7

Fueling Our Future: A Dialogue about Technology, Ethics, Public Policy, and Remedial Action is the first in a series of *Quaker Institute for the Future Pamphlets* for which there is a Series ISBN: 978-976-8142-21-4

The cover view of the Earth focusing on Africa is from Google Earth <earth.google.com>.

Philadelphia Yearly Meeting of the Religious Society of Friends Earthcare Working Group

Philadelphia Yearly Meeting of the Religious Society of Friends is the association of Quaker Meetings in Eastern Pennsylvania, Southern New Jersey, Delaware, and part of Maryland. The Earthcare Working Group's purpose is to work, within and through Philadelphia Yearly Meeting, to help bring about a transformation of the earth-human relationship so that the earth's integrity and resilience can be restored.

Quaker Institute for the Future

The mission of QIF is to advance a global future of inclusion, social justice, and ecological integrity through participatory research and discernment.

<www.QuakerInstitute.org>

Producciones de la Hamaca is dedicated to:
—Celebration and documentation of Earth and all her inhabitants,
—Restoration and conservation of Earth's natural resources,
—Creative expression of the sacredness of Earth and Spirit.

Contents

Preface	4
Energy Decisions and Friends Testimonies	9
Spiritual Guidance and Friends Testimonies	10
The Problem: Climate Change	12
Visions of Our Energy Future	15
What Can Be Done to Reach Carbon Emission Goals?	18
The Electricity Situation	21
Energy Sources for Electricity Generation	24
The Coal Situation	27
The Nuclear Energy Situation	32
The Transportation Situation	45
The Biofuels Situation	48
Financing our Energy Future	51
Ethics of Right Relationship and Energy Adaptation	54
Ethical Choices among the Energy Options	56
Endnotes	60
Bibliography	62

Preface

The climate crises has now become a climate emergency. The widespread deterioration of systems that allow human settlement to flourish is a scientifically measurable and empirically observable reality.

Scientific information about the disruption of the climate by our high-energy-comsuming civilization amounts to a new "revelation" about the human-earth relationship. New moral and ethical concerns require new thinking about how we relate to the earth and to each other.

Under the weight of this reality, people within various religious communities are struggling to understand and reformulate the ethics of their world-view. Like those in many other faith communities, members of the Religious Society of Friends (Quakers) are pondering the ethical implications of various energy technologies with regard to climate change.

In 2007 the Earthcare Working Group of Philadelphia Yearly Meeting (PYM) and Quaker Institute for the Future (QIF) began a series of consultations for discernment on three energy technologies—nuclear, coal, and biofuels. These three technologies were chosen for study because they are being promoted by various interests as essential to our energy future, but are the most ethically conflicted options before us.

Sharp divisions of opinion and advocacy have emerged among Quakers on the proposed resurgence of nuclear power. For example, Friends Committee on National Legislation (FCNL) is on record as opposing the building of new nuclear power plants, as is Quaker Earthcare Witness. Other Friends insist that a rapid development of nuclear power is desperately needed to replace coal in order to reduce CO_2 emissions. They build a strong ethical component into their argument.

Many people are not sure what to think about the promotion of "clean coal" and biofuels, both of which involve important technical and ethical considerations that require deep scrutiny. FCNL has not taken a position on "clean" coal, coal gasification, or the development of biofuels in relation to energy policy and climate change, although the ethical considerations are significant and the public policy framework is rapidly evolving.

This study was undertaken to provide information and informed analysis of these energy options, and to help those with differing and conflicting points of view about these technologies to have a better understanding and respect for one another's views. Framed by Quaker values and testimonies, consultations for discernment were held at Harrisburg (PA) Monthly Meeting on April 14, 2007, at Friends Center in Philadelphia on June 7, 2007, and at Pendle Hill (Wallingford PA) August 21 –24, 2008.

The aim of the discernment process reported here was not to create a unified voice or to advocate a particular position with regard to the energy options under scrutiny. The aim, in addition to laying out a certain amount of technical, economic, and social information, was to bring forward the assumptions behind the various positions that people take with regard to these energy technologies and the policies around them.

Fueling Our Future is about advancing dialogue, not closing the argument. As will be clear in what follows, participants in the process were not neutral. We were in close agreement on some aspects of our discernment and divided on others—most notably nuclear power. We agreed on a set of framing principles that we advance as a context for further dialogue. We hope that others will find them to be a useful contribution to ethical decision making in this complex and conflicted arena.

We have been asked why we did not focus more attention on the most hopeful energy options—non-carbon renewables. Our answer is that we found little disagreement or ethical quandary about most renewables, and there is a vast literature singing their praises and advancing their economics. The problem occurs with whether, and to what extent, nuclear, coal, and biofuels should play a continuing part in our energy support system, which is the focus of our discernment process and this report.

The impetus for this study came from the Abolish Nuclear Weapons Working Group of Philadelphia Yearly Meeting and their concern with the linkage between the resurgence of nuclear power and nuclear weapons. As the idea for a discernment process emerged it became clear that this resurgence should be explored in the context of energy options and climate change, and in connection with FCNL's linkage of preventing deadly conflict and energy policy. The planning and conducting of the consultations was handled by the Earthcare Working Group of PYM, and by the Circles of Discernment Program of Quaker Institute for the Future.

We wish to thank the Chace Fund of Philadelphia Yearly Meeting, Elaine Emmi, and the Friends Testimonies and Economics Project for financial support, and Harrisburg Monthly Meeting for hosting the first consultation. We thank all the participants who generously and professionally entered into the consultations with a spirit of discernment and who shared our sense of urgency in grappling with the ethical contexts of these energy options. We especially thank Judy Lumb for the way she joined the process and employed her writing and editorial skills in the construction of *Fueling Our Future*. From the stack of tapes that recorded every session and the accumulation of notes that were folded into the record, she has crafted a document that distills the discernment process into an accessible report. We are grateful to Charles Blanchard, Phil Emmi, Geoff Garver, Patricia McBee, Robert McGahey, Marilyn Roper, Shelley Tanenbaum, and Marianna Wood for very helpful reviews that significantly improved *Fueling Our Future*.

Ed Dreby and Keith Helmuth, Coordinators

Presenters at Harrisburg Friends Meeting
April 14, 2007

Baird Brown (Partner, Ballard Spahr Andrews and Ingersoll, LLP)

Keith Helmuth (Secretary of the Board of Trustees, Quaker Institute for the Future)

Robert McKinstry Jr. (Goddard Professor, Pennsylvania State University, School of Forest Resources)

Mark Myers (Member of the Board of Trustees, Quaker Institute for the Future, Chairperson of the Board Trustees, Earlham College and Earlham School of Religion)

Marianna Wood (Associate Professor of Ecology and Biological and Allied Health Sciences at Bloomsburg University)

Members of Harrisburg Friends Meeting and Philadelphia Yearly Meeting's Earthcare Working Group were also in attendance.

Participants at Friends Center
June, 2007

Baird Brown (Ballard Spahr Andrews and Ingersoll)

Ed Dreby (Earthcare Working Group, PYM)

Hal Feiveson (Co-Director of Science and Energy Security Program, Princeton University)

Keith Helmuth (Quaker Institute for the Future)

Hollister Knowlton (Quaker Earthcare Witness, FCNL)

Margaret Mansfield (Friends Committee on National Legislation)

Robert McKinstry Jr (Pennsylvania State University, School of Forest Resources)

Mary Burton Risely (Friends Committee on National Legislation)

Liz Robinson (Director of Energy Coordinating Agency, Philadelphia)

Dan Seeger (Quaker Institute for the Future)

**Editorial Committee at Pendle Hill
August 21-24, 2008**

Keith Helmuth (Quaker Institute for the Future)

Judy Lumb (Editor, *Producciones de la Hamaca, Quaker Eco-Bulletin)*

Robert McGahey (North Carolina Interfaith Power and Light, Quaker Earthcare Witness)

Mark Myers (Quaker Institute for the Future)

Dan Seeger (Quaker Institute for the Future)

Marianna Word (Bloomsburg University)

Mary Burton Risely (FCNL, participating via email)

Energy Decisions and Friends Testimonies

"More clearly than ever before, Friends recognize that the critical and interconnected issues of energy and environment relate not only to climate change, but also to war, military spending, the nation's budget, and the capacity of the United States to meet domestic human needs and invest in the well-being of vulnerable populations abroad. We are called to redefine the human and national security of the United States to include freedom from deadly conflict, freedom from abuse of power, assurance of basic human needs, and protection of the Earth's commons—the air, land, and water on which all life depends."

—*Friends Committee on National Legislation*[1]

The purpose of this study document is to provide Friends and others with the information needed to make decisions on public policies to advocate for our energy future. We want to build broad awareness and inspire Friends to action, so we strive to present this information in easily understood forms. We cannot foresee the future. The future will evolve differently than we imagine, but we need processes in place to deal with whatever comes. Everything about our energy future and the coming re-adaptation requires deep thought and prayerful consideration.

The challenge we face is that we are considering choices between future energy alternatives and the uses of our natural environment that do not appear fully adequate to meet the world's needs and may cause harm to the environment of future generations.

In spiritual discernment, Quakers call on another body of knowledge, God's truth as revealed by the Inner Light and Continuing Revelation. Our challenge is to hold these technical and spiritual bodies of knowledge together in such a way that a creative synthesis results from scientific truths merged with God's truths.

As a community of seekers we are challenged by the technical nature of the choices we must make and by the ambiguities and dilemmas represented. It is hard for Quakers to accept the risks of imperfect choices.

Quakerism is a highly personal religion with all having equal access to the truths revealed. It is a religion that has a strong resistance to the authority of a few. Yet, in understanding these choices, we must depend on the abstractions and advices of a few authorities, most of whom are unknown to us.

Spiritual Guidance and Friends Testimonies
Role of Science and Religion

The following description of the energy options is merely a summary of information that is well known. We have nothing new to offer in terms of the technology, or the facts concerning the technology. What Friends have to offer is a process for making decisions, an approach which considers not only the facts as we know them, but the larger truths, that reside in the ethical understandings of human solidarity and the integrity of Creation—God's truth.

We are struggling with the ethical implications of energy production and use. We are struggling with what energy options and energy policies Friends should apply and advocate. A purely technocratic approach is inadequate within the context of discernment as practiced by Friends. It is likewise problematic to couch this whole discussion in an intellectual and analytic framework. A fully rounded discernment process must hold in tension knowledge developed by the sciences, and moral intuition developed by religious and cultural guidance.

Bringing objective knowledge and moral intuition into dialogue has led Friends to conflicting conclusions on energy options. The clarification of assumptions from which reasoning and discernment on energy options proceeds can be helpful. If, for example, the combined weight of technical information and moral intuition makes the expansion of nuclear power seem unwise to many Friends, it can be seen that those who are led to contrary judgment are equally serious in their discernment.

Science is based solely on evidence, so a scientist cannot know more than the evidence presents. But when Friends are meeting in discernment as in the preparation of this document, there is a sense of relationship that has the potential for wisdom beyond the scientific facts. Critical change in ecologically sound adaptation will occur when the convergence of knowledge, understanding, and moral intuition reaches a "tipping point" in our society, and social, religious, and political leadership adequate to the task emerges.

Although we do not all share the same view of energy options, we agree that the assumptions from which we reason should be clear and transparent. Through the Quaker process of discernment, and guided by Friends' testimonies we have converged on this set of framing principles that should be considered in all decisions on energy technologies and energy policies.

Framing Principles

1) For a sustainable future, carbon emissions must be in equilibrium with the capacity of the earth to absorb carbon.

2) Carbon emissions should be decreased as soon as possible to prevent catastrophic effects of climate change.

3) Full cost accounting must include net energy benefit and life cycle cost of each option considered.

4) Decisions made today must not put burdens on future generations who have no part in making these decisions.

5) Decisions that we make must not transfer the burdens onto the poor and disadvantaged or onto other areas of the world.

6) Decision making on energy options must be inclusive, fair, equitable, grass-roots driven, and consensus based.

7) We must seek solutions that scale to the needs of all people on earth.

8) We must seek solutions that preserve the rest of the biosphere beyond humans.

Among those participating in this study, there are areas of settled conviction and agreement and other areas of controversy and disunity. All agreed on the following:

- The climate change crisis is urgent and requires immediate action to prevent catastrophic consequences.
- Right sharing of energy resources and risks must be a major consideration in the development of energy policy.

- Meaningful goals with effective enforcement are needed as soon as possible to reduce GHG to levels that will prevent harm to future generations.
- We must support and encourage energy efficiency.
- We must call for serious reduction of consumption, and provide leadership for the change in thinking and habits this requires.
- We must commit to energy efficiency and sustainability in all our Friends buildings.
- We must support and encourage distributed generation—practical electric service options at local and regional levels.
- We must support and encourage the development and use of renewable energy sources.
- There is no such thing as "clean" coal, so we must minimize the burning of coal as much and as soon as possible.
- Use of fossil fuels is not sustainable and must be replaced as soon as possible.
- Biofuel production must not compete with food production.
- Public transportation and passenger rail service must be expanded.

Areas of controversy and disunity center around the question of whether a resurgence of nuclear energy is necessary as an interim measure before we can get to the point where all energy for human activities is generated by sustainable methods from renewable sources with carbon emissions in equilibrium with carbon absorption.

The Problem: Climate Change

Since the advent of agriculture, human history tells the story of repeated environmental degradation, but in earlier eras low-density population and available land often made it possible to move on to unspoiled areas. That kind of human adaptation is long gone. We are now coming up against limits of many kinds, including the limit of climate stability in relation to greenhouse gas (GHG) emissions. We cannot move away from climate change that is affecting all parts of the

Earth, and which results from greenhouse gas emissions from human-related sources:

1) Buildings where we live, work, and shop (26%),
2) Generation of the electricity we use (41%), and
3) Transportation of ourselves and the goods we use (33%).

This document focuses on the impact upon climate change from the generation of electricity and from transportation, but does not address the impact upon climate change attributable to buildings. More information on buildings can be obtained from two Friends organizations that have undertaken major renovation projects. The FCNL building in Washington, D.C., has been renovated into a model demonstration green building, which acts as a lobbying entity itself. Friends Center in Philadelphia is in the midst of a similar renovation.

Carbon Dioxide

Carbon dioxide is the predominant GHG, accounting for 84 percent of the human-related emissions (*Table 1, p. 14*). The realities of our situation are as follows:

- Over the past half-million years, the average level of atmospheric carbon dioxide (CO_2) ranged from 180 parts per million (ppm) in the coldest to 280 ppm in the warmest times.

- After the last glacial period, it took 6,000 years for CO_2 levels to increase 80 ppm.

- In 2006, the level of CO_2 was 386 ppm, about 100 ppm higher than in the late 1800s.

- Reducing the level to 350 ppm CO_2 would avoid really dangerous change.[2]

- Holding the rise to 550 ppm CO_2 is probably the best that can be done, according to some global economists.

- Other global economists emphasize the economic costs of not acting with urgency to reduce CO_2.

- Business as usual could take us to 900 ppm by 2100.

Other Greenhouse Gases

Table 1: Greenhouse Gas Emissions from the U.S.

Year	CO_2	CH_4	N Oxides	Other[a]
1990	5,069[b]	606	383	90
1995	5,394	599	396	106
2006	5,983	555	368	148
Percent[c]	85%	8%	5%	2%

[a]Other greenhouse gases include hydrofluorocarbons, perfluorocarbons, and sulfur hexaflurorides.
[b]Greenhouse gas emissions are presented as teragrams of carbon dioxide equivalents, which takes into consideration the global warming potential of each gas.
[c]2006: Percent of Total
Source: *Inventory of U.S. Greenhouse Gas Emissions and Sinks: 1990-2006*, U.S. Environmental Protection Agency, April 15, 2008.

Consequences of Climate Change

This is not the first warming event the earth has experienced, but previous events occurred over much greater time frames that allowed time for adaptation. The current situation is changing so fast that the models developed to forecast the rate of change in climate change factors are now falling behind the actual rate of change.

Arctic ice, Greenland ice cover and Antarctic ice shelves are all giving way much earlier than predicted, and at a continuously accelerating rate of disappearance. This means that the current forecasts of both the rate and height of sea level rise could be significantly under-estimated. A significant rise in ocean level has the potential to massively disrupt coastal urban complexes—the epicenters of modern civilization. A six-meter rise in sea level will displace 200 million people world wide, 11 million in the U.S.

The collapse of the Greenland glacial structures is especially ominous with respect to altering the salinity balance of the North Atlantic Ocean. The effect could block and end the Gulf Stream flow and, thus, change the climate of northern Europe, and the north Atlantic in general, from temperate to sub-Arctic. Recent data indicate the Gulf Stream is already slowing down.

The pace and consequences of climate change are tricky to predict because climate is a nonlinear dynamic system with enormous inertia and numerous feedback mechanisms. The thermal inertia of the oceans will warm the earth an additional 0.6 degrees Celsius, even without further carbon emissions.

Most known feedback mechanisms are self reinforcing. For example, with the collapse of the Arctic summer sea ice, the polar region looses reflectivity, absorbs more sunlight, heats near surface water, warms polar atmospheric circulation and hastens Greenland ice sheet collapse. This process could add a further warming of 0.3 degrees Celsius.

Other feedback mechanisms are counterbalancing. One possibly important counterbalancing mechanism has to do with aerosol particles. Aerosol air pollution particles emitted by burning coal mask a fractional part of the warming potential of greenhouse gases. Scientists now have begun to account for the cooling effect of these particles.[3]

Facing the Moral Implications

The costs and benefits of our current energy consumption are inequitably distributed:

- Those who contribute most will be least harmed.
- Those who contribute least will be most harmed.
- The least vulnerable will be the best able to pay for re-adaptation.
- The most vulnerable will be the least able to pay for re-adaptation.

Who has the right to use the atmosphere as a carbon sink and in what amounts? The United States with five percent of the world's population contributes 25 percent of world carbon emissions. What do our testimonies as Friends require of us?

Visions of Our Energy Future

The Quaker Institute for the Future was founded on the conviction that Friends testimonies, witness, and public policy work could be strengthened by deeper analysis and clearer articulation of the human prospect within the ecological context. The context of energy production and use within human settlements worldwide—technical, ecological, economic, and ethical—is central to our future energy decisions. Policy options are fraught with assumptions, often unspoken, hidden, and shielded from critical analysis. One goal of this study is to bring those assumptions into the energy futures discourse.

Many folks, not having thought much about energy futures in these terms, are now increasingly confronted with the realities of a deteriorating human situation and are wondering what to think, wondering what assumptions about the human prospect make sense, and what they mean for both personal and societal adaptation.

Currently, we are on an unsustainable "energy flight" based on the use of fossil fuels, which are not renewable. There are four visions of our energy future.

1) Continuing the Flight: Technology will solve the problems.

Some folks think that getting the technology right will enable the species, or at least the wealthy part of it, to keep going on the current energy flight indefinitely. This modernist high-tech vision is that technology of all sorts, including expanded nuclear power, carbon capture and storage, coal, wind, solar, and biofuels of all sorts will allow us to maintain our present rate of growth in both population and energy use. This vision assumes that fusion power, and other yet-to-be-discovered energy technologies will eventually allow even further growth.

In the modernist high-tech vision of the future, demographers envision a planet of megalopolises of 100 million. Other species do not appear in these scenarios, nor do viable ecosystems. Such a world would necessarily exceed arable land resources, requiring feeding all but the most wealthy with artificial food. It is a strange world, assuming that humans can live essentially alone, being able to generate any features and resources exceeding those commoditized from the remnant of the biosphere.

2) Soft landing: Efficiency, conservation and renewable energy will solve the problems.

Most folks who have thought seriously about the human trajectory and our ecological realities, want to see our current energy flight come in for a soft landing, a climb down from energy use based on fossil sunlight to energy use based on current sunlight and bio-productivity. The most widely shared alternative vision among Friends assumes that we can do it all with increased efficiency, conservation, and renewable energy sources.

This vision implicitly includes an upper-middle to upper-class economic base, with sufficient private money to capitalize technologies that are expensive, like photovoltaic panels and small, relatively inefficient windmills, supplemented by government support. This scenario assumes that developed industrial nations are morally obligated to vastly reduce their energy use, but it does not address how the poor, especially the industrial poor in the developing world, will manage this transition.

3) Medium-hard Landing: Global economic depression

Some folks think that a global depression would retard economic development sufficiently to prevent runaway climate change. But in such circumstances desperate people would use anything they could for fuel and electricity, returning to plentiful coal and available biomass. Even if obtainable petroleum reserves receded to the vanishing point, liquefied coal and synthetic fuels are not difficult to produce. Both increased nuclear capacity and a rapid build-up of renewable energy sources and the infrastructure to support them would suffer greatly. A global economic collapse would also exacerbate current growing inequalities between rich and poor, and economic injustice is a driver of a number of factors which undermine ecosystem stability.

A growing number of sober analysts now think the chances of the soft landing option have vanished. A medium-hard landing would mean widespread suffering of environmental refugees, exacerbated by famine, war, and pestilence, but enough political structure, both democratic and autocratic, would survive for many societies to muddle through.

4) Hard Landing

The most dire vision is that effective global action on climate change was delayed too long. The result would be a hard landing with a cascade of system breakdowns requiring rapid, and largely improvised, re-adaptation under deteriorating and highly volatile circumstances. This civilizational disaster would bring mass human migrations seeking habitable land, wars over dwindling resources, natural and technological systems overwhelmed by the magnitude and pace of climate change. Unless there is very rapid, globally coordinated action on climate change, this scenario and the assumptions behind it will become increasingly a part of the energy future landscape.

What Can Be Done to Reach Carbon Emission Goals?

There is no silver bullet. We must take a portfolio approach, selecting from the various options what will likely work in any given jurisdiction.[4] We must create redundancy and overlapping strengths in systems to insure resilient and sustainable operations. There is no perfect system; all have risks, but the risk of climate change dwarfs all other risks.

Because the United States is one of the biggest sources of carbon emissions and the goal is to provide information for Friends in the U.S., this pamphlet is limited to the U.S. response to the climate change crisis. This pamphlet focuses on the most controversial of the various fuel options for generation of electricity—coal, nuclear energy and biofuels—which will be discussed in more detail beginning on page 28. The following is a summary of some other strategies to reduce carbon emissions.

Efficiency and Conservation

Following the current business-as-usual trajectory, the demand for electricity would double by 2050, but we could do something about that by working at the level of conservation and increasing efficiency.

1) Decrease the demand for energy.

2) Generate electricity more efficiently.

3) Prevent losses in transmission of electricity.

Different ways of thinking about energy and energy use can change habits. During the last 20 years California has had both population increase and economic growth without an increase in electricity use.

Geothermal Heating and Cooling

Geothermal heating and cooling makes use of the constant (52 – 55 °F) temperature below the earth's surface. Pipes can be laid and liquid circulated through buildings for cooling in the summer and heating in the winter. This greatly reduces the need for fossil fuels for heating and air-conditioning.

Methane Reduction by Altering Cow Feeding

Cow burps account for 23 percent of methane emissions.[5] Cows emit methane because they are fed a corn-rich diet. Corn is rich in sugars that ferment in the cow's stomach. The resulting methane gas is burped out into the atmosphere. But cows do not need the sugars; they need the protein. If, in feed preparation, the sugars were separated from the protein, the cows would get what they need without producing and emitting methane. The sugars could then be turned into biofuel for a double win. The increasing trends toward vegetarian diets and reduction of beef consumption can also be an important factor in methane reduction from agricultural sources.

Use of Coal

Coal could be used more efficiently and with reduced carbon emissions, but "clean" coal is merely an advertising campaign. Various methods of sequestering carbon emissions from coal plants have been suggested, but none has reached the pilot project stage. Coal will be discussed in detail later. (*p. 28*)

Carbon Sequestering

In addition to all the other carbon sequestering schemes that are under discussion, we should remember that building with wood is a carbon sequestering action. Wood and fiber-based materials could be used to replace much of the metal, plastic, and inorganic composites that are now widely used in building. The significance of forestry resources and the sustainable care of forest lands could be greatly advanced.

Smart Growth

Smart growth would have us living and working with goods, services, recreation, and cultural amenities all clustered for easy, low-energy access. Urban cores would be renewed with urban and suburban agriculture. Goods and services would be produced and marketed regionally with shortened transport supply lines. Public transportation would be broadly available, along with pedestrian and bicycle friendly roads. Building codes would require green building standards for all new and renovated buildings.

Societal and Lifestyle Factors

It is not just about technology; it is also about societal dynamics and the resulting lifestyles. Policies that create financial incentives and disincentives are effective in changing individual and societal behavior. Peer influence works. Non-material rewards, relationship benefits, and ethical satisfactions can reduce consumerist behavior.

Population

Human impact on the earth is complicated by increasing numbers of humans. Demographers make a range of projections, the mid-range of which is that the global population will level off around nine billion by 2050. Can the earth support that many people? Probably not, but population control is never benign. We have the example of China, which used very oppressive methods to control their population. In contrast is the example of Mexico where persuasion by a media campaign was quite successful.

But numbers of humans are not the whole story. Decreases in fertility are associated with increases in consumption. Because of our over-consumption, the addition of one person in a developed country has much more impact on the earth than one in a developing country.

International Protocols and the U.S. Response

International agreements are extremely important to solve the climate change crisis. The problem is unequal representation among nations and peoples. We need to equalize representation from all countries of the world and to expand the process to embrace native cultures and spiritual traditions. International agreements must be binding with effective, loophole-free enforcement.

At the Earth Summit in 1992 the United Nations Framework Convention on Climate Change was formed. Five years later the Kyoto Protocol set targets to reduce greenhouse gas emissions for 37 industrialized countries and the European Union to an average of five percent below 1990 levels. The Kyoto Protocol was adopted in Kyoto in December, 1997, and entered into force in February 2005.

While the U.S. signed the Kyoto Protocol in 1997, the U.S. Senate passed a unanimous resolution that no treaties would be ratified that did not also include binding targets for developing countries, so the U.S. never ratified the Kyoto Protocol. By 2008, 183 other nations had ratified the Protocol.

In December 2007, the 13th Conference of the Parties (COP-13) adopted the Bali Action Plan, the goal of which is a new global climate change agreement to be executed in 2009 at the Copenhagen COP-15. At COP-14 held in Poznan, Poland, in December, 2008, some progress was made on financial mechanisms for funding of adaptation and mitigation efforts in developing countries supported by developed countries. Again, the U.S. inhibited the process of international agreement on climate change. President Barack Obama has promised that under his administration the U.S. will engage vigorously in the global process to solve the climate crisis.

While no progress was made at the federal level toward reducing GHG emissions under the Bush Administration, states and many municipalities led the way. Fourteen states have set serious goals for carbon emission reductions—50% to 85% on various timetables ranging from 2040 to 2100. Arizona has instituted 49 policy measures, New Mexico, 69. Vermont is the most stringent with a commitment to a 50% reduction by 2020.

The sooner the better is the rule for reductions. Large reductions early in the timetable are more effective than the same level on a longer timeline. The processes have been largely stakeholder driven, bottom up, facilitated, and run by consensus. They have produced over 300 technical and social mechanisms at the state level. A wave of learning and innovation transfer took effect as new jurisdictions came on stream and benefited from the accumulating technical and social mechanisms that have been put into practice.

The Electricity Situation

Electricity generation creates 41% of U.S. carbon dioxide emissions and electricity demand is expanding. The current generating system is running short on capacity, which leads to chronic emergency, decreasing reliability, economic and social disruption.

System Components and Terms

Electricity systems consist of the energy source, generation facilities, the high voltage transmission system over long distances and the low voltage local delivery to the consumer. The background demand for electricity is called "base load." "Peak load" occurs during times of greatest demand. Winter and summer peak loads occur for increased heating and air-conditioning needs. Reserve generating capacity must be available to be brought on line quickly to meet peak load or to cover for a generation facility failure or shutdown. Natural gas, hydro, coal and biofuels can be used for both base load and peak load electricity. Nuclear energy sources are appropriate only for base load because of the time required to bring increased capacity on line. Wind and solar generation are intermittent sources and require a higher level of reserve power.

The whole system has to work together all the time to provide service. Electricity is not a commodity that can be measured out, transported around, and stored for future use. It is a condition that has to be everywhere present at the same time for the system to work.

The idea that electricity is a commodity is a fundamental fallacy that has resulted in the unsustainable system we now have. Electricity was priced using economies of scale—the more you produce the cheaper it gets, so the more you use the better rate you will be given. The incentive works in the wrong direction from an ecological point of view. The charge for electricity service should be an infrastructure cost, and not for what happens when you flip a switch.[6]

Regulation

One hundred years ago societal management was conducted within a variety of resilient information and communication venues with high levels of redundancy. The electricity industry developed as regional, regulated monopolies with rate structures set by public interest regulators. The industry in the U.S. has now moved through deregulation to market competition among independent producers.

Now modern economic and social life is invested in the uninterrupted operation of a single, increasingly fragile electricity system, with no adequate alternative. The ability to manage our society depends almost totally on an unfailing electricity system.

Financing

In the era of regulated monopolies, financing was straightforward since all critical factors were known or highly predictable. Electric utilities were considered the safest investments available. Financing is now much more complex and problematic. Long-term contracts are needed but fluctuations in market conditions mitigate against confident long-term planning.

Many of the renewable generating projects are small scale, some very small, and the big investment banks generally won't touch anything less than $50 million. Renewable energy technologies are, therefore, harder to finance. Tax deductions and government grants for wind and solar installations can help support small-scale renewable electricity generation.

Demand Management

Demand management involves reduction of peak load and overall reduction of demand for electricity. Conservation and efficiency in energy use can net considerable reduction in demand for electricity.

The new telemetering technology in passive appliances (freezers and refrigerators) can turn them off and back on in coordination with lower peak load demand. Wide-spread application of this and other efficiency measures are needed.

Transmission System

Today's alternating-current (AC) transmission system needs a major upgrade for our energy future. It is in poor shape and at the limit of its capacity. Since the northeast U.S. blackout in the mid 1960s, an increasing number of grid failures have occurred at various scales. The ice storm in the northeastern part of the continent in 1998 that literally crushed the grid is a notable example.[7] Hurricane Katrina over New Orleans is another story of cascading infrastructure collapse.

The AC transmission system is very inefficient as there is considerable loss during long distance transmission of AC electricity. Recent studies indicate that transmission of high voltage direct current (HVDC) is more efficient over long distances. Integrated systems where HVDC is used for long distance and converted to AC for local transmission are being developed in Texas to allow massive development of wind energy.[8]

Distributed generation provides an alternative to supplement the centralized distribution system. Distributed generation means a wide variety of onsite, neighborhood, municipal, and regional generating technologies, all with short transmission networks. The best place to put effort is in support for local, self-organizing, off the grid, distributed generation—the more management from the bottom up, the more resilient and wiser the outcome.

Table 2: Energy sources for electricity generated in the U.S.

Source	Generated[a]	Percent[b]
Coal	1,990,926	47%
Petroleum	64,364	2%
Natural Gas[c]	829,104	20%
Nuclear	787,219	19%
Hydroelectric[d]	289,246	7%
Biomass	54,758	1%
Wind	26,589	<1%
Solar/Photovoltaic	14,568	<1%
Geothermal	507	<1%
Other[e]	139,174	3%
Total	4,064.538	100%

[a] Electricity generated in 2006 in megawatt-hours.
[b] Percent of total electricity generated in 2006.
[c] Includes other manufactured and waste gases, all derived from fossil fuels
[d] Includes pumped storage.
[e] Includes batteries, chemicals, hydrogen, pitch, purchased steam, sulfur, and tire-derived fuel.
Source: *Annual Energy Review 2007*, Report No. DOE/EIA-0384 June 23, 2008, Energy Information Administration, Department of Energy.

Energy Sources for Electricity Generation

Current Sources

Fossil fuels accounted for 69% of electricity generation in 2006. (*Table 2, p. 24*) Some non-carbon-based renewable energy sources include wind, solar, and hydro generated by rivers, tides, or ocean waves. There are technological limitations for each, but some will be significantly overcome by intensive development.

Wind Power

Wind power has several major advantages. After manufacture and installation, wind turbines are totally GHG-emission free. There are subsidies currently available for development of wind energy in the U.S. and in 2008 it is economically feasible. Denmark is currently supplying 20 percent of their total electricity by wind. Wind has the smallest landprint for the energy produced.[9]

The major downside is that wind generation is intermittent and fluctuating, so capacity must be stored during peak generation times to be used for peak demand. The best sites are in the central area of the U.S., long distances from population centers where electrical energy is needed, so wind power has the same transmission issues as the present central generation and long distance grid system. Development of HVDC transmission systems would significantly increase the efficiency of transmission.

There are aesthetic objections to wind farms in some places; and they can be dangerous for migrating birds, although that problem is being addressed.

Solar Power

Solar power is GHG-emission free. The only impacts are the embodied energy from manufacture and land use on large solar farms. Solar-generation of electricity is also intermittent but not so much so as wind power. Large-scale system use is restricted to high sun regions such as the southwestern U.S. Several studies have indicated that a large investment in solar power installations and high-voltage direct current (HVDC) transmissions systems could significantly reduce the need for electricity generated by GHG-emitting fuel sources or nuclear energy.[10]

Solar technology is evolving rapidly and the cost is dropping. Paint-on solar electricity generating coating will make any suitable surface a production site. This technology has the potential to replace the current chip-array systems and transform the electricity environments of human settlement in many parts of the world to completely carbon-free, decentralized, distributed generation. Solar energy can also be used to generate heat (solar thermal), which may have even more potential in the future generation of electricity.

Geothermal Electricity Generation

There are areas such as The Geysers area of California where very hot water heated by molten rock under the earth's surface can be pumped to the surface and used to generate electricity. This source of energy for electricity is GHG-emission free and constant, so it is suitable for base-load electricity generation. There are still some technological issues to be resolved before this source can be developed completely, but many suitable sites are found in the western U.S.[11]

Hydroelectric Power

Hydroelectric generation is GHG-emission-free. Large dams that create reservoirs are good for reserve capacity, but are problematic because of the environmental damage caused by large areas being inundated. Most appropriate sites for large reservoirs have already been used, but there are old small dam sites on eastern U.S. rivers that could be reclaimed for low-head hydro operations or "run-of-the-river" dams.

Hydroelectric power is especially good for managing peak demand. During low demand times, water can be pumped to a high reservoir to be released in high demand times without any start-up period, so the system responds by generating electricity within 15–30 seconds.

Tidal generation has potential for development, especially in northern latitudes where there are large tidal changes. Since tides are predictable, this technology could be developed for base-load power.

Open ocean waves can also be tapped for hydro-electric energy. Because waves are created by wind, the same issue with intermittency applies as with wind energy.

Nuclear Power

Other non-carbon-based fuels include nuclear energy. Pennsylvania currently sources 40% of its electricity use with nuclear generation. France is up to 80% of its electricity generated by nuclear. Many other jurisdictions worldwide are steadily increasing their nuclear generating capacity. The pros and cons of nuclear energy are discussed in detail later. (*p. 33*)

Carbon-based Renewable Energy Sources

Biofuels from sugar cane, corn, switchgrass, prairie grasses, small diameter hardwoods, used cooking oil, and other current waste products could either be burned to generate electricity or used to generate fuels for vehicles. Biofuels are a good base-load electricity energy source. There is a wide range of potential fuels that can be used at competitive rates. But biofuels are not GHG-free and may even emit more carbon than fossil fuel sources. Biofuels will be discussed in detail later. (*p. 48*)

Natural Gas

Natural gas is methane that occurs as a fossil gas, associated with oil or coal, or in non-associated gas fields. Natural gas is used for residential cooking and heating, industrial processes, electricity generation, and hydrogen generation for fuel cells. Burning natural gas produces less pollution and more energy than coal, but it is a fossil fuel that is not renewable and will be depleted.

Methane

Methane from landfill emissions being used to fire gas-powered generators is a big improvement over flare burning to dispose of the gas. Methane has the same low post-combustion emissions as natural gas. Manure is an agricultural source of methane. Coal plants can be converted to use poultry manure mixed with wood chips. Using methane from manure for fuel in gas-fired generators solves three big problems—odor, GHG emission, and water pollution from runoff to streams. This is a win all the way around and should be done at every available site.

Cogeneration

Cogeneration occurs when the waste products of one operation are used for another purpose. Heat generated by electricity-generating plants can be used for heating buildings. In another example, the production of sugar from sugar cane creates considerable plant waste material, which can be converted into biofuel to be burned for generation of electricity, making the whole operation cyclical and self-contained. This is an efficient distributed energy system where the electricity is generated on-site, eliminating loss from transmission and the need to dispose of the waste materials.

The Coal Situation

Coal is the dirtiest fuel we use. Clearly, coal as an energy source is the biggest contributor to GHG emissions. In 2008 nearly half of the electricity in U.S. was generated by 1,493 coal-fired generators with a total capacity of 335,830 megawatts. Over 100 new coal plants are currently proposed, but in the past few years many of those proposed have been cancelled. In 2002 the Department of Energy projected that 36,000 megawatts of new coal-generated electrical capacity would be added by 2007, but only 4,500 megawatts were added. Concern for climate change is one of the reasons for this trend.[12]

As the world is coming to the end of oil, coal is still available and cheap. It is commonly reported that coal reserves in the U.S. could last 150 years, assuming some increase in usage. However, this calculation assumes the energy output from high quality coal with high energy quotients. But high quality coal is nearly mined out and the remaining coal deposits have medium to low energy quotients. It will, therefore, take a far greater tonnage of the remaining coal to provide the same energy as high quality coal. This overlooked discrepancy in the calculation of resource reserves dramatically shortens the timeline for coal depletion and will greatly increase the estimated future cost of energy generated from coal. World-wide electricity demand could be satisfied for a mere 50 years with all the coal available.

From a pricing perspective, the amount of coal remaining is less important than the timing of peak coal production. As access to readily mined reserves declines, the rate of production slows. After the production rate peaks, demand will rapidly outpace supply and price per ton will accelerate. Several studies identify a worldwide coal production peak in the mid 2020s.

The U.S., China, India, and Brazil all have economies based on coal reserves. Until some recent off-shore drilling, Brazil had no oil resources available. China is building an average of one new coal plant per week. In one year alone, 2005, they built the total capacity that the UK has built since the industrial revolution began.

Coal is dirty at every stage. Extraction of coal from the earth is the most dangerous work in the U.S. Farmers have the distinction of the largest total number killed on the job, but coal miners have the largest number of worker deaths *per capita*.

One can see the scars of strip mining and mountaintop removal in western Pennsylvania and slag heaps from deep mining in eastern Pennsylvania. Water leaches sulfuric acid from the slag, which pollutes the streams and kills most living things. Remediation efforts have reclaimed some areas that have been worked recently, but there has been no remediation of areas worked in the past, so there are still large wastelands.

The fly ash resulting from the burning of coal in power plants accumulates and contains heavy metals and carcinogens. The spill of more than a billion gallons of coal ash slurry from the Tennessee Valley Authority coal power plant at Kingston, Tennessee, on December 22, 2008, shows that coal is dirty even after burned. The slurry covered all the houses in the area and flowed into the Emory and Clinch Rivers, which are tributaries of the Tennessee River.

Coal is a sedimentary rock of carbon, oxygen, and hydrogen that carries sulfur and nitrogen impurities, as well as uranium and thorium. More radioactive contamination occurs around coal plants than around nuclear plants.

One concern about the combustion of bituminous coal, the most abundant form of coal available now, is the emission of sulfur dioxide and nitrogen oxides, which are air pollutants with multiple harmful effects. They cause acid rain, contribute to smog, and form particles that are harmful to breathe. The acid rain problem took 30 years to address and caused tremendous damage in the meantime. Today, emissions of sulfur and nitrogen oxides have been greatly reduced by scrubbing the effluent, reducing the temperature in combustion chambers, and employing other control measures. A great deal has been done, which gives power to the industry's argument that future issues can be addressed. However, these solutions did not reduce CO_2 emissions, which is the big problem now.

Sequestering Carbon

Carbon sequestration is capturing the carbon dioxide and pumping it into the ground for a long period of time, which must be done under higher than atmospheric pressure to reduce the volume. The best bedrocks for sequestering are not necessarily next to coal-fired plants, so some kind of extensive piping network would have to be created, possibly to the same extent as our Interstate highway network. Neither the U.S. nor the European Union governments have put much investment into the technology for cleaning up coal.

There is no way to know that the gas will stay in the ground. When do you call sequestering successful? Suppose the gas remains underground 20 years, but not 50 years. CO_2 is an odorless gas, which is heavier than oxygen. If suddenly released in large amounts, it can smother all life dependent upon oxygen.

This has already happened at Lake Nyos, Cameroon, Africa, which sits on top of an inactive volcano that releases CO_2 into the water. In August of 1986 cloud of CO_2 was suddenly released from the lake and killed 1,700 people and 3,500 livestock in nearby villages.[13] This raises a liability issue. Who owns the CO_2 and whose responsibility is it if a cloud of CO_2 escapes and smothers a whole community?

In January 2008, the Bush Administration withdrew funds from FutureGen, a major carbon capture and sequestration project which was planned in Illinois, saying the cancellation was due to cost overruns. There is a smaller pilot project in Britain and one successful example in the North Sea where they are pumping carbon dioxide into areas where they have extracted oil.

There are more desirable approaches to sequestration than pumping gas into the ground, but they are economically uncertain. For example, carbon dioxide-based minerals are some of the most common in the crust of the earth, such as limestone. Concrete is carbon dioxide-based. Sequestering in stable mineral products seems a better option than pumping gases into the ground.

Carbon dioxide can be used to grow algae. The effluent smoke is bubbled through tubes seeded with algae, which allows the growth of algae in the tubes. The algae can then be put into a landfill, which is a more reliable form of sequestration than as compressed carbon dioxide. Or, the algae can be pressed to harvest oil, which can be burned as bio-diesel. The Massachusetts Institute of Technology has a pilot project where algae are used to scrub the effluent from their power plant.

Clean Coal

The term "clean coal" came out of advertising by the coal industry. "Clean" coal is an oxymoron. Mining is dangerous, dirty, and damaging. Each year workers die in mine accidents. Mountaintop removal creates terrible impacts on the surrounding environment. Until we wean ourselves from dirty and dangerous energy sources, we are ethically compromised.

Gasification of Coal

In addition to carbon sequestering, "clean" coal technology includes coal gasification, a refinery process that turns coal into a gas that can be burned to produce electricity. The process takes out all the mercury, sulfur, and particulates, so the gas is much cleaner burning than coal. Residual emissions from burning gasified coal are on par with natural gas. But there are serious concerns about the impacts of the refinery process and the costs of gasification that must also be included in the overall assessment of this technology.[14]

Liquid Fuel from Coal

Coal gasification technology can be used with chemical catalysts to make synthetic liquid fuel which could replace Diesel or jet fuel. But even with carbon capture and sequestration involved in the refinery process, this option will emit even more carbon than the petroleum-based fuels it replaces. Thus, development of liquid coal is a dead end and should not be pursued.

Political and Economic Considerations of Coal

The political reality is that coal will be used for the foreseeable future. If new coal technologies are not developed, the old plants will continue to be used. The industry will argue that they should go ahead and build new plants, assuming that sequestering technology will be developed later.

New technologies could increase efficiency and reduce emissions. While there has been no reduction in the amount of carbon dioxide emitted, newer coal technology has increased the efficiency from 40 percent to nearly 60 percent, so the amount of carbon dioxide emitted per kilowatt-hour has decreased. Electricity generation produces heat, which can be used for heating if a city is close to the power plant. Most coal plants could be adapted to biofuels.

U.S. policy has been to keep both energy and food prices far below market value. That is why people are feeling so oppressed now because we have been so used to cheap food and energy. In other countries people have been paying more for energy and food for a long time.

If the sequestering of carbon dioxide is developed, it will add significantly to the cost. It is estimated that to put in all the remediations would add at least 50% to the price of electricity produced with coal, but we will not know the actual cost until the projects are completed. It would be ethically irresponsible to go forward with coal without sequestering. If we could say that the carbon dioxide had to be sequestered, it might make coal so much more expensive that we could begin to divert the political process that now favors coal. However, investment in new coal technology detracts from investment in renewable sources which are sustainable.

We could use natural gas, but that is more expensive, so the economics work in favor of coal. We must move away as quickly as we can from fossil fuels. There is no argument that fossil fuels have viability in the long term. Fossil fuels change the balance because they are out of equilibrium. We are releasing carbon that was drawn out of the system millions of years ago. Any use of fossil fuels is temporary. They will be depleted.

If carbon is priced high enough in cap and trade systems, that will send us in the right direction. But the operation of the market is not an automatic solution. Markets are not free, they are controlled. The market is often ethically oriented only in the first "split second" of its response to an opening. Almost immediately distortions, power moves, and control factors come into play. There are important areas of decision making that are outside the influence of the market, such as land use planning.

We are at the historical moment where a new social contract can be negotiated, with carbon taxes and appropriate pricing of food and energy. The old social contract cannot be supported any more. We recognize that government policies must shift, but we were lacking the political leadership. With the Obama Administration in place there is great hope that this will change.

The Nuclear Energy Situation

In 2008 one-fifth of U.S. electricity was generated by 104 nuclear power generators with a total capacity of 105,585 megawatts. No new nuclear power plants have been constructed since the accident at Three Mile Island in 1979. Thus, all nuclear plants in the U.S. are more than 30 years old. The original licenses were for 40 years, but some of these licenses have been extended for another 20 years.[15]

This means that over the next few decades the current nuclear energy plants will be nearing the end of their lifetimes. Just to keep nuclear energy generating the same amount of electricity will require new plants. This reality and concerns over carbon emissions from coal-generated electricity has led to a resurgence of interest in nuclear power. In 2007, applications for eight new nuclear power reactors were submitted to the Nuclear Regulatory Commission (NRC), in 2008 applications for 18 new reactors were submitted, and applications for nine more are expected in 2009 for a total of 34 new reactors at 23 locations.[16]

The proposed reactors include two NRC-certified and three proposed designs. 1) Advanced Boiled Water Reactor (ABWR) made by General Electric was certified by the NRC in 1997. 2) AP1000 made by Westinghouse (Toshiba) was certified by the NRC in 2006. 3) Evolutionary Power Reactor (EPR) was designed by Avela (France) and Siemens (Germany). Two EPR reactors are now under construction in Finland and France. 4) Economic Simplified Boiling Water Reactor (ESBWR) made by General Electric is advanced over the ABWR with more safety features. 5) Advanced Pressurized Water Reactor (US-APWR) is made by Mitsubishi.

The Bottleneck

Seven billion tons of carbon are now emitted annually worldwide. Business as usual will take us to 14 billion by 2050. The most dire predictions envision runaway climate change that will doom civilization, and probably assure early extinction of the human species. The probability of this vision grows higher every year that we delay mitigative actions.

Given the reality that climate change effects are happening even faster than the models have predicted, it is clear that we must stop burning coal unless carbon capture and storage can quickly become commercially feasible.

Our goal is sustainable, carbon-neutral energy sources and technologies for electricity, transportation, heating and cooling—a neo-agrarian future where this level of sustainability has been attained, and any non-sustainable methods would be unthinkable, including both burning fossil fuels and extracting ever-diminishing ores from which fissionable materials can be generated.

In order to get to this sustainable, carbon-neutral future, we must pass through a bottleneck brought on by our heavy dependence on fossil fuels and inadequate development of renewable energy sources.

The question is whether we can get through this bottleneck without the resurgence of nuclear energy. Since the participants in the consultations did not come to unity on the answer to this question, arguments are presented for (PRO) and against (CON) in parallel below.

PRO:
Assumptions of Those Promoting Nuclear Energy

- The benefits of nuclear energy outweigh the risks.
- Using nuclear energy allows a nation to get closer to carbon neutrality.
- Nuclear power has a smaller landprint than most other energy sources.
- Nuclear energy enjoys substantial economies of scale, so the larger the installation the smaller the unit price for the electricity produced.
- Under the current regulatory regime, nuclear power is cost-effective over the lifetime of an installation, roughly 60 years.
- Nuclear power is needed to meet base-load electricity demand.

CON:
Assumptions of Those Opposed to Nuclear Energy

- No levels of radioactive emissions are harmless.
- Overall life-cycle risks of nuclear energy are unacceptable.
- While nuclear reactor safety has been improved, the incalculable consequence of catastrophic breakdown is a fact that can never be overcome.
- Nuclear energy cannot be separated from nuclear weapons proliferation.
- The problem of nuclear waste disposal has not been solved.
- Uranium supplies are dwindling and will be increasingly more expensive to mine.
- Nuclear power requires centralized, corporate-controlled energy production and distribution.

- Development of new nuclear energy plants will require huge investment and subsidies that could be better used to develop renewable energy sources.
- The decade or longer time lag between proposal and operation means that nuclear energy plants can not resolve current needs nor address early-phase bottleneck issues.

PRO: Nuclear Reactor Safety

Three Mile Island and then Chernobyl gave the world two contrasting examples of nuclear accidents. The accident at Chernobyl was catastrophic. The flawed Chernobyl-style reactor design is still used in Russia and the Eastern European heirs of the Soviet empire. Fears of catastrophic results of another nuclear accident are warranted for these reactors, which need to be replaced or modified to render them acceptably safe.

But Three Mile Island, though costly in terms of lost power, repair and reconfiguration, was a success story because the container vessel did its job. No significant levels of radiation were emitted. The Three Mile Island incident set the nuclear industry in the U.S. back for a generation, but it led to tremendous improvements in reactor safety. Operator safety training, including simulated near-accident conditions, has been hugely increased, with operators spending a significant portion of their work hours in such simulations. Moreover, reactors have been redesigned with high level safeguards against sabotage and human error, with key shutdown switches operating by gravity. The number of valves has been reduced, and backups to critical plumbing systems providing water to damp any potential runaway reaction have been built. If one system fails, the other automatically goes into operation.

Reactor safety studies show that events requiring shutting down the reactor have decreased 100-fold since the Three Mile Island incident. The "accident sequence precursor" index: was 7.3 in 1980, 1.2 in 1990, and less than 0.1 in 2000 (per 7000 hours of operation). This means that efficiency is way up, with nuclear power plants currently working at 90% of capacity, versus 70% or less in the past.[17] The main reason to shut down a nuclear reactor now is to refuel. The likelihood of a nuclear reactor accident happening in the U.S., France, Japan, the UK, or Scandinavia is almost nil.

A U.S. company has run a successful test in Moscow with a Chernobyl-style reactor retrofitted with a thorium-burning insert. Thorium is even more abundant than uranium, and both the reaction and the waste generated are much safer than burning uranium. The thorium reaction is driven by a proton gun. To stop the reaction, you simply turn off the gun; there is no possibility of the reaction going critical and causing a meltdown.[18] The waste products have much lower half-lives than those of uranium once-through reactors.[17] It is imperative that the Chernobyl-style reactors be replaced with reactors with containment vessels, or retrofitted with thorium reactors as soon as possible. Industry executives in the U.S. are unwilling to assume the huge costs of replacing our uranium-powered system with thorium, but in India, which does not have a developed industry, the world's first thorium-powered reactor went on line last year.

CON: Nuclear Reactor Safety

A recent study from the Union of Concerned Scientists (UCS) finds that the U.S. has strong nuclear energy safety standards, but serious safety issues arise at nuclear plants because the standards are not adequately enforced by the Nuclear Regulatory Commission (NRC), partly due to an inadequate budget and partly due to a poor "safety culture." For example, in 2001 NRC inspectors were so concerned about safety problems they submitted an order for immediate shutdown of the reactor at Davis-Besse for detailed inspection. But NRC managers were persuaded by the plant owners to ignore these concerns. When the inspections were finally done six months later, they found a large hole in the reactor head. An accident more serious than Three Mile Island was narrowly averted.[19]

In the past, public hearings have stimulated increased attention to safety issues in the nuclear industry. In order to facilitate the early licensing of new nuclear power plants, new Nuclear Regulatory Commission procedures have eliminated public hearings from the process of licensing new nuclear power plants. In another mechanism to decrease the time required for approval of new nuclear power plants, the NRC now licenses both construction and operations at the same time.

The UCS report also expresses concerns about protection of nuclear power plants against sabotage and terrorist attack. It is recommended that more realistic scenarios be tested and that the oversight of nuclear power plant security be moved to the Department of Homeland Security.

Of the designs for proposed nuclear reactors, the UCS report states that only the ESBWR design has the potential to be safer and more secure than current reactors. No commercial insurance companies will provide coverage for nuclear power plants, so without the federal Price-Anderson Act, which provides insurance and limits the liability for the nuclear power companies, nuclear power plants could not operate. All of which underscores the recognition that the consequences of nuclear accidents are literally incalculable. They are off the scale of not only acceptable but of even possible risk calculation.

PRO: Nuclear Waste Disposal

The current practice for nuclear waste disposal in the U.S. involves two steps. The first is to immerse the spent fuel rods in water on-site. After a number of years, these are removed from the water and put into dry lead casings, where they are stored on-site indefinitely. Disposal of high-level wastes remains a huge challenge.

In 2002 Congress authorized funds for planning and developent of a long-term waste storage site on land that has already been used for testing and development of nuclear weapons at Yucca Mountain in Nevada. It would be funded largely from the nuclear waste fund, supplied by the tax of 0.1 cent per kilowatt-hour on electricity from nuclear reactors.

Geologists familiar with the study and the specific site are convinced that it is as strong a plan as is reasonably required to store wastes for 10,000 years, though Congress has demanded further study of what happens to the material after that period. The design has gone through many reviews and has been strengthened to provide triple fail-safe leakage lead containment systems. If one system leaks, another contains it, with yet another system in place to contain any potential further seepage. At the end of 10,000 years, there is some radioactive release into groundwater and local radiation pollution that would affect the local area.

Reports surfaced several years ago about potential falsification of water infiltration figures by hydrologists, but an independent review of the data commissioned by DOE, reporting to Congress in February 2006, confirmed its reliability.

Nuclear waste disposal is not a technical problem, but a political one. While we benefit from the energy generated by nuclear power plants, no one wants the waste stored in their backyard. Delays over building a long-term waste storage site are increasing the costs. Industry analysts say that a second site would be needed shortly after Yucca Mountain is opened, especially with the proposed increase in capacity.

CON: Nuclear Waste Disposal

No satisfactory solution to the nuclear waste disposal problem has been found. The Yucca Mountain underground storage site has not yet been approved, but even if it is approved, the cross-country transportation of nuclear waste is a significant public hazard. Dispersed, on-site storage is now the default system. The timeline for this storage is so far beyond predictable control that it cannot be placed within any kind of ethical risk assessment and decision-making framework.

Under the Bush Administration, the Department of Energy proposed the Global Nuclear Energy Partnership aimed at reprocessing and recycling nuclear waste. The proposal was to build one central facility where all the spent fuel from the entire country can be reprocessed. The reprocessing would separate out cesium for above ground storage for 300 to 400 years until it cools sufficiently to be placed safely underground. The remaining waste would then be burned in a fast neutron reactor. It would require multiple burn cycles over a period of perhaps 100 years to reduce this material to a condition in which it can be safely stored at Yucca Mountain. The Global Nuclear Energy Partnership was not supported by Congress.[20]

The risk factor in the release and dispersion of even a small amount of plutonium is biologically catastrophic on a potentially large scale. To expand the use of a technology that harbors this risk is ethically unsupportable. This is not a risk that can be parsed. This is a total risk.

PRO: Nuclear Weapons Proliferation

Nuclear weapons are a major problem in the world and should be eliminated altogether. But eliminating nuclear energy will not eliminate nuclear weapons. Nuclear energy can be completely separated from nuclear weapons proliferation. We cannot let the weapons question determine the answer to the power question.

Nuclear bombs cannot be fashioned from materials used in conventional nuclear power stations; the uranium grades are simply too low. The waste from nuclear power plants is very highly radioactive, so it cannot be easily stolen. There is no known case of plutonium being diverted from a power reactor for weapons.

However, when spent fuel is reprocessed for re-use in breeder reactors there is a danger of materials being diverted for nuclear weapons use. These reactors end up with far less highly reactive waste, which is why the French can put all of their waste in one room, but it can also be more easily stolen. The U.S. outlawed breeder reactors in the 1970s because of these security concerns.

Other technical means for reducing plutonium-related proliferation risks include using thorium or self-contained reactors in smaller countries. South Africa is interested in exporting pebble bed reactors, which would be returned at the end of their lifetime for waste processing.

Another potential association with weapons-grade nuclear materials involves research reactors that have nothing to do with nuclear energy, but produce radionuclides for medical testing and research. These reactors have the ability to enrich uranium, not only for medically useful radioisotopes, but weapons-grade material as well. The reactors under suspicion or surveillance in North Korea and Iran come under this category.

Other cultures have taken a very different view of nuclear power. For example, the people of India have had no ethical hesitancy in embracing nuclear power on a wide scale. Nuclear power will most likely become a prominent source of electricity worldwide. In the U.S. there is hesitancy about it. Is there a nuclear hangover in American culture with respect to the morality of having twice used nuclear bombs on civilian populations? Nuclear weapons and nuclear energy were entwined within the U.S. Energy Department budget. Does this association taint objective and even ethical considerations around utilizing nuclear power? If other jurisdictions, especially Third World countries, are willing to run the risks of using nuclear power, how, with respect to equity (shared risk for the sake of the common good), can opposition to domestic nuclear power development be ethically justified if coal is the alternative?

CON: Nuclear Weapons Proliferation

At the initial use, the fuel for a nuclear reactor cannot be used for nuclear weapons production. However, when the spent fuel is reprocessed, plutonium is produced that could be stolen and used for nuclear weapons. So, the extent to which nuclear energy is linked to nuclear weapons production depends upon whether reprocessing is involved.

However, every country that has acquired nuclear weapons in the last twenty-five years—India, Pakistan, North Korea—did so through acquiring the materials and technology for nuclear energy production. Every kilogram of highly enriched uranium and plutonium on the planet increases the risk of further nuclear weapons proliferation.

PRO: Nuclear Energy Fuel Supply

Uranium is a finite, depletable resource like fossil fuels. Secondary supplies are also available, for example, from decommissioned nuclear weapons, but these supplies are limited. In 2005 only 60 percent of the nuclear fuel used was from a primary, mined source. The nuclear power industry expects that new technology will make more efficient nuclear plants. With re-processing, increased enrichment, and the use of thorium, which is more abundant than uranium, the nuclear fuel supply will be adequate for a long as 200 years, enough to get us through the bottleneck.[21]

CON: Nuclear Energy Fuel Supply

Uranium is a non-renewable resource that is already dwindling. Like fossil fuels, as the more easily accessible ores are depleted, the cost of mining uranium will increase. In 2005 it was estimated that on earth there are about 3.5 million tons of uranium that are economically exploitable. At that time the 438 nuclear power reactors in operation were using 67,000 tons per year, which amounts to a 50-year supply at current usage rates. By 2007 the estimate was raised to over 5.5 million tons of uranium due to reassessment of existing sites rather than identification of new sites. It was noted that the cost of exploration has increased because they must drill deeper, so the cost of exploitation will also increase.[22]

PRO: Health Effects of Nuclear Energy

By comparison to the massive carbon overload which is a known, suicidal poisoning, the danger of releasing deadly but localized radiation from commercial nuclear operations constitutes a lesser risk. The health effects of increased nuclear power development should be compared with the health effects of continued global warming. A slightly higher rate of cancer deaths at a slightly younger age from exposure to a slightly higher level of radioactivity is not even comparable to the damage risk of irreversible climate change.[23]

CON: Health Effects of Nuclear Energy

There is no safe level of exposure to radioactivity. All exposure is toxic and cumulative, and effects continue for decades.

There are 520 abandoned uranium mines in the "checkerboard" area of northwestern New Mexico, some lands owned by the Navajo tribe, some by the state of New Mexico and some by private parties. Uranium mining has contaminated soil and ground water, indicated by elevated rates of cancers and birth defects in both people and animals. The 1979 Church Rock spill of uranium mill tailings into a stream bed continues to cause illness and death to Navajo sheepherders and their animals. The Ambrosia Lake mine and mill are just two of several Superfund sites never remediated. The Navajo Nation, the Environmental Protection Agency, Bureau of Indian Affairs, Department of Energy, Indian Health Service and the Nuclear Regulatory Commission are now working on a five-year plan to address the health and environmental impacts of uranium contamination in this region. The corporations once profiting from the uranium industry have gone away by now. Shouldn't the costs of this remediation and the ill health of those affected be added to the cost of the nuclear power plant fuel produced?

PRO: Urgency of Climate Change Crisis

If runaway climate warming is already underway, we are in for big changes. We should be considering what kind of re-adaptation is implied. We should be saying, "We have a problem. We need a process for figuring out what to do about it." Climate change due to carbon emissions is potentially so catastrophic that we should not rule out any technology in advance, but keep the whole toolkit, which includes nuclear energy.

The argument for nuclear power is for one among multiple energy sources as a bridge to get through the bottleneck, which, if attainable at all, will probably take 100-150 years. We are in a short timeframe for effective action—before 2020. It typically takes 12 to 15 years for a new technology to be developed. We do not have enough time to depend on this eventuality.

Nuclear fission is the most efficient way to provide base-load power. In terms of technologies on the table and ready to go, it is also the safest, since the other candidate for base load power is coal, which even when burned efficiently with scrubbing of the effluent, still releases unacceptable levels of CO_2, not to mention mercury, SO_2, and levels of radiation which exceed many times that which is allowable by the Nuclear Regulatory Commission.

France has a far smaller *per capita* carbon footprint than the U.S. due, in part, to its 80% reliance on nuclear power to generate electricity. In its construction, mining, and transportation operations, nuclear power plants have about the same carbon emissions as coal plants. However, the operation of a nuclear plant generates electricity without emitting carbon. Because the capacity to generate electricity from each nuclear plant is very high, replacement of coal plants with nuclear plants in the U.S. would considerably reduce our carbon emissions.

The claims for the optimistic scenario that we can do it all with increased efficiency, conservation, and renewable energy sources are based upon projections from within the renewable energy industries themselves. If we want to take new nuclear plants out of consideration, we need the confidence that sustainable sources of energy could take up the huge hole left by removing coal from our fuel sources.

No study accepted by the Inter-Governmental Panel on Climate Change in their 2006 overview predicts less than an increase of 2°C, and only 6 of 177 studies get us below 2.5°C. All of these studies included an expansion of nuclear power as part of climate mitigation. Any increase in nuclear energy replacing coal-generated electricity would be an advance. Even 5% to 10% would help.

CON: Urgency of Climate Change Crisis

Even with the NRC's new procedures designed to fast-track licensing of new nuclear plants, it will be at least ten years before any of these come on line. But every new report from the Intergovernmental Panel on Climate Change paints a more urgent picture of the climate crisis. Reductions in carbon emissions are needed now. Focusing on nuclear energy as the main solution to the crisis, as the Bush Administration appears to have done with funding priorities, ignores the urgency of the crisis. Energy conservation and improvements in energy efficiency can bring reductions more immediately. Wind and solar technologies are sufficiently developed and simpler to execute so they can come on line sooner than new nuclear power plants.

PRO: Nuclear Energy and the Peace Testimony

The Hindu/Buddhist *Ahimsa* concept is to do no harm. But the ethic of doing no harm stands relative to the greatest harm we have all perpetrated, both to ourselves and to the earth with the excess of carbon emissions for the past 200 years. Each of us witnesses from within the modern industrial system. None of us is pure, nor outside it. This applies as well to those of us who prematurely assume that we can do it all with renewables, while having to fall back on coal and nuclear-sourced base-load power. All we can do is strive to do the least harm.

CON: Nuclear Energy and the Peace Testimony

A nuclear power station inevitably means men and women with guns to guard the dangerous nuclear technologies inside. An increasingly nuclear-powered United States means growing the national security state with its secretive intelligence services and militarization of our financial and societal resources. More nuclear power plants mean more and more men and women with guns. This increasing militarization of economic, social, and political life is contrary to Friends' work for societal development that leads to a culture of peace.

PRO: Nuclear Energy and the Simplicity Testimony

The moral issue is not nuclear power, but how we care for the web of Creation. It is a question of stewardship in an era when our numbers are overwhelming the earth, multiplied by an extravagant lifestyle. Consider our freezers, clothes dryers, air-conditioning, electric heat, multiple-car families, the car itself. We live in a world deeply, perhaps fatally, compromised by our industrial choices. So much that we take for granted in modern life is anything but simple, both materially and spiritually. As long as we remain involved in a complex, high-demand, electricity-based way of life, we have little moral basis for rejecting nuclear power, especially if it can be substituted for Creation-damaging fossil fuels.

CON: Nuclear Energy and the Simplicity Testimony

Nuclear power requires a political economy of centralization that will subvert the full and rapid development of ecologically sound, non-polluting, renewable energy technologies. Compare this economy to the decentralized power production made possible by renewable sources, wind and solar, especially.

The Quaker testimony of simplicity should lead us to recommend strongly that conservation and efficiency are our first tools in solving the global climate crisis. Conservation is the fastest response we can make to the climate change crisis. We can lead the way in promoting conservation, frugality and the strengthening of local communities that decentralizing electric power production and distribution will bring.

PRO: Nuclear Energy and the Equality Testimony

The equality testimony, rightly understood, extends to all life forms and seeks to allow them their place in the biosphere. Insofar as nuclear power can help reduce the biotic catastrophe of climate change, it supports the equality testimony. By providing the possibility for lighting and heating old inefficient buildings, nuclear power would serve as a safety net for the poor who cannot afford to renovate their homes or install expensive renewable energy technology.

CON: Nuclear Power and the Equality Testimony

Nuclear power has always outsourced its risks to U.S. taxpayers through subsidies like the Price-Anderson Act, which limits the nuclear industry liability, and the promise that our government will take care of nuclear power's nearly eternally hazardous radioactive wastes. Loan guarantees are needed for construction of new nuclear power plants because no private investors will take the risks. Nuclear power plants cannot make a profit without these subsidies.

Provision of more nuclear fuel implies opening or reopening uranium mining on a vast scale. Expanding nuclear power also means shipping uranium and spent fuel across the land and water, exposing the planet to unnecessary risks of global radioactive contamination. To avoid diverting the risk of danger and damage to the disadvantaged or others who did not choose it, we must apply a vital precautionary principle. We must now ask before we commit ourselves to any course of action involving technology or the introduction of new chemicals into our planetary ecosystem, "has it been proven to be beneficial and safe?"

PRO: Nuclear Energy and the Integrity Testimony

Integrity means being willing to be honest with ourselves, to accept new information, even if it means upsetting settled values and world views. Nuclear power is an awkward compromise that requires us to honestly face its potential for helping to mitigate catastrophic climate change.

Because a relatively small amount of uranium fuel produces a large amount of energy, nuclear power is less disruptive of the natural environment in terms of mining, emissions, and landprint, than is coal. Nuclear energy is better than coal energy in maintaining the human-earth community as it struggles to survive the disruptions climate change will bring. The overall moral issue is not nuclear power *per se*, but how we care for the web of Creation. What counts, morally speaking, is our stewardship in an era when human numbers and extravagant lifestyles are overwhelming Gaia. We live in a world deeply, perhaps fatally, compromised by our industrial choices.

CON: Nuclear Energy and Integrity Testimony

Nuclear energy violates the integrity of creation by messing with the atomic nucleus by splitting the atom. This is another manifestation of our usurping power beyond our ability to use it wisely. Extensive mining of any kind upsets the integrity of the earth, and long term nuclear wastes increase the local mutative capacity. To accept this overstepping of boundaries as a means to soften the transition to a more sustainable world is to contravene integrity.

The Transportation Situation

Our transportation system that carries passengers (5.6 million passenger-miles in 2006) and freight (3 million ton-miles in 2006) is responsible for one-third of the U.S. carbon emissions. The U.S. love affair with the private automobile leads us to a most inefficient use of energy, which provides an opportunity for considerable savings in carbon emissions. In 2006, the latest year for which data are available, 88 percent of passenger-miles were on U.S. highways, while 11 percent were on airlines, and one percent of passenger-miles were via public transit or rail. Commuters report in a 2005 survey that 63 percent of the time they drove alone in an automobile, 22 percent they drove with others in the car, seven percent of the time they commuted by public transportation, and seven percent of the time they walked or rode a bicycle.[24]

Rail transportation uses half or less of the energy that would be used by rubber tires on a highway to carry the same weight, whether passengers or freight. New magnetic levitation high speed trains in Japan and Germany do not even have the friction between the rail and the wheels. The U.S. is far behind other developed countries in regard to development of high speed passenger rail systems. While air and highway travel have tremendous government subsidies, rail transportation has very little U.S. government support.[25]

The U.S. missed another opportunity to reduce GHG by a 30-year delay in requiring increase of vehicle fuel economy. In the oil crisis of the 1970s, Congress established "Corporate Average Fuel Economy" (CAFE) standards, but sports utility and trucks were exempted. Because of strong opposition from the U.S. automobile corporations, no increases in CAFE standards were made until 2007. Under the Bush Administration, the Environmental Protection Agency (EPA) denied California's request to enforce higher GHG emission standards. In his first week in office, President Obama directed the Department of Transportation to begin implementing the new CAFÉ standards and directed the EPA to review California's application to regulate GHGs.

Electric Cars

Electric cars were the first cars, before the era of the internal combustion engine. In 1900, electric cars sold more than all steam and gasoline automobiles combined. They worked well as city cars or neighborhood vehicles and are still used in limited areas such as golf courses and campuses. But once a highway system was developed, cars with longer ranges were desired, so the internal combustion gasoline engines were preferred.

In 1990 the California Air Resource Board (CARB) mandated that increasing percentages of vehicles with zero emissions be sold in California, beginning in 1998, so electric cars again entered the market. However, the U.S. corporations were less than enthusiastic and neither advertised nor supported their electric vehicles. Auto dealers sued the CARB in Federal Court, arguing that it was technically and financially impossible to fulfill the mandate. CARB backed down on the early standards and extended the time to meet the goals. Even though the market was proven for them, the companies withdrew the electric cars.

Hybrid Cars

Hybrid cars combine an internal combustion engine with an electric motor. The first production models were the Toyota Prius in 1997 and the Honda Insight in 1999. They seamlessly change back and forth between the two motors, use braking to recharge the battery, and get more than 50 miles to the gallon. In April 2008, sales of Prius passed one million worldwide, 591,000 in North America.

Hydrogen Fuel Cell Cars

Hydrogen fuel cells combine hydrogen and oxygen to produce electricity and water, the only output of these clean cars. A battery is used to facilitate and help manage the electrical power of the drive motor. The battery is recharged using the braking system as in the hybrid cars. Honda has provided a few of its new FCX Clarity models for lease in Southern California.

The problem with hydrogen fuel cell cars is the source and containment of hydrogen. Honda has set up a few refueling stations in Southern California where they produce the hydrogen from natural gas. They are working on a home energy system that would provide heat and electricity for the home, as well as hydrogen for the vehicle. This would be an improvement over the carbon emissions from the home using commercial electricity and fuel for heating, but for future sustainability a renewable source of hydrogen must be developed.

As we know from the 1937 Hindenberg dirigible accident, hydrogen is extremely flammable and dangerous. Several safety features are built into the Honda model and have been successfully tested, but safety remains a concern.

The Biofuels Situation

All internal combustion engine fuels are biofuels. Petroleum is a biofuel from ancient storage—a kind of geological savings account. Ethanol and bio-diesel are biofuels from current photosynthetic income. By August 2008 corn ethanol and soy bio-diesel were the most common commercial biofuels used in the U.S. The two current biofuels in the U.S., corn and soy, are food crops, so biofuels are competing with food. So much corn is now going for ethanol that Iowa may soon have to import corn for feedstock. In 2008 it is very clear that is a real problem.

Production and Use of Corn Ethanol

Corn ethanol has become widely used as an additive in gasoline as an oxygenator, replacing the additive methyl tertiary butyl ether (MTBE). MTBE reduced the tailpipe emissions and boosted octane ratings, but turned out to have serious health risks, contaminated ground water supplies, and proved almost impossible to cleanse from the environment. Different research protocols for determining net energy gain or loss in the ethanol production cycle have yielded different results. Recent advances in crop yields and biofuel production efficiencies accounted for the observed 25 percent net energy gain, whereas earlier analyses sometimes showed a net energy loss.[26]

A 10 percent blend of corn ethanol with gasoline is now common. Blends at 85 percent are becoming increasingly available. Conversion of engines to running on ethanol is fairly simple and inexpensive. Corn ethanol production in the U.S. in 2005 was 12 billion gallons, approximately one percent of the petrol fuel consumption budget. If all the corn grown in the U.S. were turned into ethanol it would supply 12 percent of the U.S. transportation needs.

Meanwhile, government, agribusiness, and farmers have gone for maximum corn ethanol production. Corn ethanol production receives a $0.50 per gallon Federal subsidy, which is in addition to the Federal subsidy that corn production alone receives if corn prices are low, since farm subsidies guarantee a minimum price. When corn prices rose in summer 2008, the federal subsidy dropped out for that season.

Production and Use of Soy Bio-diesel

Current production and use of soy bio-diesel is 6 percent of the U.S. soybean crop. Soy bio-diesel is a pressed oil seed product requiring no fermentation, as does ethanol. Soy bio-diesel can be produced with far less inputs than ethanol and produces less GHG. Nitrogen fertilizer used for corn production is 70 times that for soy per net energy unit from biofuels, and phosphorous fertilizer used for corn is 13 times that for soy. Pesticide application per net energy unit for corn is ten times that for soy. Soy bio-diesel has a net energy gain of 93 percent.

Other Fuel Stock Plants

Hulless barley requires less processing than corn. Switch grass is a contender. *Miscanthus*, a hybrid from China and Japan, is a possibility because it will not grow where you don't want it to grow because it will not naturally reseed itself.

Harvesting and processing the naturally occurring, or reestablished, mixture of prairie plants is under study. This low impact, high diversity option is very attractive for its natural regeneration, and because the continuing root structure maintains a large carbon sequestering factor. They are perennials and hold onto nitrogen so they don't need fertilization. They can grow on marginal land that wouldn't be used for food crops.

With each of these grasses, the problem is breaking down the cellulose so the fermentation to ethanol can go on. There has been an effort for some time to solve this problem, but not much has been accomplished. They could be burned right now, so the most efficient current use for transportation is to burn them for electricity and use electric vehicles.

The 2008 energy legislation added a subsidy of $1.01/gallon for production of cellulosic ethanol and $45/ton subsidy to growers of biomass feed stocks. This is new for cellulosic, in addition to the subsidies already in place for corn.

Canola and oil palm are also being cultivated for conversion to biofuels. The growth of the oil palm market is a direct threat to topical forests. Although some jurisdictions have tried to restrict forest destruction (mainly burning) for creation of oil palm plantations, this conversion, under pressure of the market, continues to occur. Sugarcane, sugar beet, and cassava have good potential for biofuel production.

Potential of Algae and Other Options

In addition to the fact that algae can be grown in potentially vast quantities in contained structures ("algae farms"), it can also be grown in situations where it first performs a scrubbing function that removes carbon from emissions. It can then be processed for its oil content and turned into a biofuel, a double win.

Methanol collection from landfills and used for local needs is another option. Closed systems such as dairies using manure for their electricity generation are other options. Waste oil products such as fryer oil from restaurants can be used as a stock for biofuels.

Climate Change Implications of Biofuels

Carbon-based biofuels from current photosynthetic income release less CO_2 when burned in high-efficiency internal combustion engines than does gasoline. If the cycle of use can be folded back into the farm machinery operation needed to produce the fuel stocks, a further gain will accrue. With soy bio-diesel the net energy gain is so high that a sustainable cycle from farm through the market and back to the farm can be foreseen. Corn ethanol may or may not have this option depending on how the net energy gain/loss question works out.

However, internal combustion engines burning biofuels still emit significant levels of CO_2. The fuels are renewable and not fossil fuels, but their use as an alternative fuel for vehicles does not constitute a "clean" car. It may be better to drive electric cars and use the biofuel to generate electricity with carbon capture and storage.

Ethics and Ecology of Biofuels

Biofuels raise serious issues of land-use ethics. The landprint for use of biofuels is very high. To operate all the automobiles in the U.S. on 85 percent ethanol fuel uses one million times the land needed for wind-generated electricity for the same electric car use. As fuel crops compete with food crops, food prices are rising and will hurt the poor the most. Should drivers have a claim on the land that pre-empts food production?[27]

Even if food crops are not used as biofuels, inappropriate land might pressed into use for biofuel cropping. Fuel crops intensify unsustainable monoculture practices. We should be moving toward, not away from, sustainable agriculture. Increased intensive use of mono-cropping and wearing down of soil structure and fertility comes with large scale industrialized farming. As corn prices rose in 2008, partially but not entirely due to federal mandates for minimum biofuel use, there was a substantial increase in marginal acreage in corn.

The rush to produce biofuels in vast quantities masks the need to rethink transportation systems in terms of ecologically sound adaptation and resource use footprint in general. Reducing demand should not be forgotten. Reducing demand, reducing use, is still the single biggest contribution to emission reduction that can be made. Increasing fuel efficiency standards, closing the SUV exemption gap, and expanding public transport systems are close behind.

Financing our Energy Future
Subsidies for Oil and Coal

The oil industry in the U.S. receives large direct subsidies, including lower tax rates than other industries, large deductions for costs associated with drilling and exploring for oil, and the depletion allowance, which allows sheltering of up to 20 per cent of the proceeds from a well, assuming that it will one day be depleted. This allowance applies to all mining, oil and natural gas wells, and timber. Instead of paying the cost of depleting natural resources, it amounts to a financial advantage for the industry. The result of these subsidies is that gasoline prices in the U.S. have been much lower than in the rest of the world, both developed and developing countries. Over the years these subsidies have kept the cost of fossil fuel use artificially low, making the renewable alternatives appear less economically feasible, which has inhibited the development of solar, wind and other alternatives.

Subsidies for Renewable Biofuels

Recently, biofuels have been touted as the solution to U.S. dependence upon foreign oil, so they, too, benefit from subsidies. Ethanol from corn receives a federal subsidy of $0.50 per gallon, which is in addition to the subsidy received for corn production, if the market price is below a fixed value. Cellulosic ethanol has a subsidy of $1.01 per gallon.

Liability Limits for Nuclear Energy Plants

The Price-Anderson Act limits the liability of companies operating nuclear power plants and provides liability insurance through a pooled insurance fund to which the companies contribute. Beyond the funds available in that fund, the U.S. government compensates members of the public who incur damages from any radioactive accident, no matter who is liable. This covers the operation of all nuclear reactors for testing, research and energy generation, transportation of radioactive materials, and all costs involved in any incident, such as, precautionary evacuation, incident response, investigation, settling of claims, and defense against these claims. In the 43 years of Price-Anderson protection, the nuclear insurance pools have paid $151 million for claims and the U.S. government has paid $65 million. The Price-Anderson Act was renewed in 2005 and is now in effect until 2025. Without this government-backed insurance, nuclear energy generation would not be possible.[28]

Loan Guarantees for Nuclear Energy

Even with this legal exemption from law suits, investors are skeptical about nuclear power development in the U.S., so the government provides loan guarantees for nuclear energy development. For fiscal years 2008 and 2009, the U.S. Congress appropriated $18.5 billion for loan guarantees for construction of new nuclear energy plants and another $2 billion for new uranium enrichment plants. Fourteen projects to build 21 reactors applied for loan guarantees, but the total cost for these projects is estimated to be $188 billion. The industry is asking for $122 billion in loan guarantees.[29]

Loan Guarantees for
GHG-Free Renewable Energy Sources

For many years active lobbying by the oil industry kept monies for development of renewable energy such as wind and solar to a minimum. Beginning in 2005, $2 billion was available for federal loan guarantees designed to take large new energy technology projects from pilot or demonstration level to full implementation. Pre-applications were received for 143 projects totaling $27 billion, nearly half of which were for biofuels and 16 percent for coal gasification and carbon capture and storage. The sixteen projects approved to submit a formal application were for hydrogen fuel cell development (1), electric car development (1), electricity transmission (1), solar technology (2), industry energy efficiency (2), coal gasification (3) and biofuels (6).[30]

Table 3. Loan Guarantees in U.S. FY 2009 Budget

Project	Allocation in billions
Nuclear Power Plant Construction	$18.5
Uranium Enrichment Plant Construction	$2
Coal Carbon Capture and Storage Projects	$6
Coal Gasification	$2
Renewable*	$10
Total	**$38.5**

*Renewable includes biofuels, solar, wind, wave, tidal, and geo-thermal, as well as transmission and distribution technologies.
Source: Department of Energy, "DOE Announces Loan Guarantee Applications for Nuclear Power Plant Construction." Press Release, October 2, 2008.

U.S. Government Priorities

U.S. government priorities under the Bush Administration were readily apparent from energy technology projects approved in 2006 and the 2008 appropriations for loan guarantees (*Table 3*). More than half the projects approved in 2006 were for coal and biofuels. More than half of all the funds appropriated for 2009 energy technology development were for new nuclear power plants, $20.5 billion of a total of $38.5 billion. Coal technologies were also singled out for loan guarantees, leaving all renewable technologies, transmission, and distribution projects to share the remaining $10 billion.

Tax Credits for Consumers

A tax credit for hybrid cars was in the Energy Policy Act of 2005. The credit phased out once that automobile company had sold 60,000 cars, so the program was not designed to encourage tax-payers to buy hybrid cars, but to allow other manufacturers to get into the market that Toyota and Honda had served for several years.[31]

The 2008 U.S. Emergency Economic Stabilization Act included tax credits for tax-payers making home energy efficiency improvements, installing solar or wind electricity, and installing geothermal heating and cooling.

Ethics of Right Relationship and Energy Adaptation

Future Generations and the Disadvantaged

We have a duty to future generations and to the poor and disadvantaged to reduce our carbon emissions and not take more than our fair share of the atmosphere and all other resources. By what right do the wealthy of the world continue to accumulate wealth and persist in appropriating a vastly disproportionate share of Earth's resources to their own use?

What in nature and human knowledge do we have the right to own? How do property rights get established, allocated and reallocated? Do we have the right to deplete soil fertility, to use agricultural land for biofuel production, to contaminate freshwater sources, to contaminate the air, or to dispose of nuclear wastes that will contaminate the Earth for tens of thousands of years?

The only cheap energy resources are fossil fuels, which would not be so cheap if there were full accounting and if subsidies were lifted. But as fossil fuels are depleted, oil becomes more expensive and a lesser quality of coal must be used. Energy and food become more expensive. The equity issues are huge. What will happen to low income folks? Inadequate income issues must be solved by guaranteeing a living wage for everyone and not used as a basis for energy policy.

To share resources fairly across the whole earth, carbon quotas could be *per capita* so everyone on earth would have a certain amount of carbon emissions to spend as they pleased. Since seven billion tons of carbon are now being emitted annually and there are over six billion people on earth, that quota would be roughly one ton of carbon per person per year. In 2007, the U.S. emitted 1,632 million tons of carbon, which is more than five tons per person.

Obligations to the Biosphere

Survival of life on earth depends on our ability to ally ourselves with the interdependent diversity and self-management of the biosphere. A more balanced and sustainable human presence in the biosphere requires a reduction of consumption in overdeveloped regions and a rethinking of the economic structures that support overconsumption.

The current economy depends on the conversion of Earth's capacities into commodities, and commodities into monetary wealth. The aim and goal of societal development worldwide has been and continues to be the accumulation of monetary wealth and access to consumer goods by continuing and accelerating this conversion. Already the regions of industrial development and great wealth accumulation are living on natural capital instead of renewable income. The total human draw on the natural capital has exceeded the Earth's bio-productive and bio-assimilative capacities since 1986. If the consumer economy of the U.S. is the aim of China, India, etc., the prospect for the human future is dim indeed.[32]

Some traditional societies model a socially rich and materially modest way of life in which Earth and its whole commonwealth of life are experienced as a sacred community. We must develop an alternative vision of a global economy that incorporates this understanding of the sacredness of the biological community.

Revisiting our Framing Principles

This discussion began with a set of framing principles on page 11. We now revisit the framing principles to consider our energy choices.

1) For a sustainable future, carbon emissions must be in equilibrium with the capacity of the earth to absorb carbon.

2) Carbon emissions should be decreased as soon as possible to prevent catastrophic effects of climate change.

3) Full cost accounting must include net energy benefit and life cycle cost of each option considered.

4) Decisions made today must not put burdens on future generations who have no part in making these decisions.

5) Decisions that we make must not transfer the burdens onto the poor and disadvantaged or onto other areas of the world.

6) Decision making on energy options must be inclusive, fair, equitable, grass-roots driven, and consensus based.

7) We must seek solutions that scale to the needs of all people on earth.

8) We must seek solutions that preserve the rest of the biosphere beyond humans.

Ethical Choices among the Energy Options

The post-hydrocarbon future that we move toward will require more diverse and complex sources of energy, as well as much more efficient use and conservation of energy. We are seeking solutions to a complex systems problem that involves economics, physics, chemistry, biology, geology, and political science.

Of the general options presented, coal without carbon capture and storage, and gasoline-powered vehicles violate the first framing principle because neither allows equilibrium of carbon emissions. Long-term storage of both CO_2 and nuclear wastes violates the fourth framing principle because both put the burden on future generations.

While biofuels are in equilibrium with carbon emissions because they are emitting current carbon just fixed by the plants, massive development of this option violates the fifth and eighth framing principles because the plant growth requires too much of the earth's surface, which eliminates space for other species and competes with the production of food.

Massive expansion of public transportation and re-establishment of a nationwide rail system are essential to reduce carbon emissions. Electric cars provide one good option, but the generation of electricity must be factored into the decision. Fuel cell cars are another option, but the generation of the hydrogen must be considered. Current operations use natural gas, which violates the first framing principle.

New hydroelectric dams violate the eighth framing principle because much land is inundated in the reservoirs. Some sites might be appropriate for small, run-of-the-river electricity generation that would not violate the framing principles.

That leaves geothermal, wind, solar, and other hydroelectric generation such as wave or tidal methods for generating electricity, which represented only 1 percent of the total electricity generated in the U.S. in 2006. To avoid the catastrophic impact on human settlement from climate change, major changes in carbon emissions need to be made immediately, or at least within the next ten years.

The argument for nuclear energy is based on the assumption that renewable energy options cannot fulfill energy needs for the next few decades and nuclear power is needed temporarily until the renewable technologies are developed.

The nuclear waste issue is such a problem that it seems if we make the decision to go with maximum development of nuclear power, we also need to approve reprocessing of nuclear waste. That links nuclear energy with nuclear weapons technology, which makes this option unpalatable to Friends.

Keeping nuclear power going prolongs the dependence on centralized production and a long distance distribution system that is now increasingly fragile and prone to failure. If we forego the nuclear energy option (build no new plants and run out the life cycle of current plants), and close coal plants that do not capture and safely sequester CO_2, it would force the maximum development at the most rapid rate possible of distributed generation and renewable energy technology.

All of the various options must be implemented simultaneously: population control, energy efficiency and conservation, over-developed countries drastically curtailing material consumption and energy use, massive development of solar, wind, and other renewable energy sources, implementation of a more efficient long distance transmission system, and rapid development of highly diverse distributed generation.[33]

The question is whether implementation of all of these can be done quickly enough to prevent runaway climate change and a civilizational disaster. Several prominent energy analysts and planners are convinced that solar electric energy can develop rapidly enough to end investment in nuclear and coal plants without creating a massive shortfall.[34]

Is it reasonable to assume that a nation that mobilized technology and social cooperation with the speed it did for the Second World War, cannot now, in the face of climate change, mobilize its technology and people for a rapid redevelopment and transformation of its electricity system, and its energy systems in general? At the worst, a short fall of the capacity to power all current end uses may occur, and a sloughing off of non-essential consumption would be required to meet basic service. Such an adjustment could be seen as the force of necessity moving our society in an ecologically responsible direction. This would be not unlike the food rationing of the Second World War that produced flourishing "victory gardens" over the whole country.

But would communities respond in a cooperative manner to adjust to the new realities? How do you maintain a civil society under these conditions? What if there is no emergence into equilibrium and the critical period continues in perpetuity?

Meeting the climate change emergency requires a degree of economic wisdom, social organization, information synthesis, and technological integration that greatly exceeds anything we have ever experienced. Whatever we are going to do responsibly requires a degree of human cooperation that we have never seen happen on this scale.

Even though our energy decisions have global effects, this document has focused on the U.S. alone. Since the U.S. is only 5 percent of the population but emits 25 percent of the carbon, we must target the U.S. economy and the policies that control it for leadership in carbon emissions reduction. However, our energy choices are only for the U.S. To a large extent the direction that the rest of the world takes is independent of what we do. One global contribution we could make is to invest in the development of solar, wind, and other technologies that could be exported to other countries, especially to developing countries.

In the Quaker tradition of engagement with the critical tasks of our social and economic situation, and with the wider welfare of Earth's commonwealth of life in mind, we offer this report on our energy options discernment process. Measured against the scale of the climate change and re-adaptation in prospect, it seems a small gesture. But we hope it advances greater dialogue and wider discernment among Friends, as well as prompting direct action on remediation and public policy.

Endnotes *(full citations in the Bibliograhy)*

1 Friends Committee on National Legislation sets priorities for each Congress. This quote is from the prioties for the 111th Congress <http://www.fcnl.org/priorities/priority_111th.htm>

2 Hansen, James E. *et al*, 2008.

3 Hansen, James E., 2007 and Lenton, *et al*, 2008.

4 Blanchard, Charles and Shelley Tanenbaum, 2008.

5 Environmental Protection Agency, 2008.

6 Patterson, Walt, 2007.

7 Doheny-Farina, Stephen, 2001.

8 Stovall, John, *et al*, 1987.

9 Jacobson, Mark, 2009.

10 Makhijani, Arjun, 2008; Scheer, Herman, 2007; Zweibel, Ken, et al., 2008.

11 Geothermal Technologies Program, 2008.

12 Freese, Barbara, *et al.* 2006.

13 Baxter, P.J., and Kapila, M., 1989.

14 Freese, Barbara *et al.* 2006.

15 Energy Information Administration, 2008b.

16 Nuclear Regulatory Commission, 2008.

17 Bodansky, David, 2004.

18 Bodansky, David, 2004.

19 Ronlund, Lisbeth, 2007.

20 Department of Energy, 2008c.

21 World Nuclear Association, 2005; International Atomic Energy Agency, 2007.

22 Mudd, Gavin M., and Mark Diesendorf, 2008.

23 Lovelock, James, 2006.

24 Department of Housing and Urban Development, 2006.

25 Lumb, Judy, 2008.

26 Hill, Jason, *et al* 2006.

27 Jacobson, Mark, 2009.

28 American Nuclear Society, 2005.
29 Department of Energy, 2008a.
30 Department of Energy, 2007.
31 Internal Revenue Service, 2006.
32 Simms, Andrew, 2005; Athanasiou, Tom, 1998; Chang, Ha-Joon, 2002; Laidi, Zaki, 2007; Brown, Peter G., and Geoffrey Garver, 2009; Nadeau, Robert, 2006.
33 Patterson, Walt, 2007; Zweibel, Ken, *et al.*, 2008.
34 Bradford, Travis, 2006; and Scheer, Herman, 2007.

Bibliography

(websites accessed Decembe 29, 2008)

Adger, W. Neil, Jouni Paavola, Saleemul Huq, and M. J. Mace, Eds, 2006. *Fairness in Adaptation to Climate Change.* Cambridge MA: MIT Press.

American Nuclear Society, 2005. The Price-Anderson Act. <ans.org/pi/ps/docs/ps54-bi.pdf>.

Athanasiou, Tom, 1998. *Divided Planet: The Ecology of Rich and Poor.* Athens GA: University of Georgia Press.

Ausubel, Jesse H., 2007. Renewable and nuclear heresies. *International Journal of Nuclear Governance, Economy and Ecology* 1(3): 229-243.

Baxter, P.J., and Kapila, M., 1989. Acute health impact of the gas release at Lake Nyos, Cameroon, 1986: *Journal of Volcanology and Geothermal Research* 39:265-275.

Barnes, Peter, 2006 *Who Owns the Sky* and *Capitalism 3.0: A Guide to Reclaiming the Commons.* San Francisco CA: Berrett-Koehler.

Beck, Ulrich, 1995. *Ecological Politics in an Age of Risk.* Cambridge: Polity Press.

Beck, Ulrich, 1999. *World Risk Society* – Cambridge MA: Polity Press.

Blanchard, Charles and Shelley Tanenbaum. 2008. What Can We Do about Climate Change? *Quaker Eco-Bulletin* 8(2):1-4.

Bodansky, David, 2004. *Nuclear Energy:* Principles, *Practices and Prospects*, 2nd Edition. Seattle WA: Springer.

Bradford, Travis, 2006. *Solar Revolution: The Economic Transformation of the Global Energy Industry*, Cambridge MA: MIT Press.

Brown, Peter G. and Geoffrey Garver, 2009. *Right Relationship: Building a Whole Earth Economy.* San Francisco CA: Berrett-Koehler.

Bureau of Transportation Statistics, 2005. Omnibus Household Survey, Table 7. Department of Transportation <bts.gov/publications/omnistats/volume_04_issue_01/pdf/entire.pdf>.

California Environmental Protection Agency Air Resources Board, 2003. *2003 Zero Emission Vehicle Program Changes.* Fact Sheet. <arb.ca.gov/msprog/zevprog/factsheets/2003zevchanges.pdf>

Chang, Ha-Joon, 2002. *Kicking Away the Ladder: Development Strategy in Historical Perspective.* London: Anthem Press.

Crosby, Alfred W., 2006. *Children of the Sun: A History of Humanity's Unappeasable Appetite for Energy,* New York NY: W.W. Norton.

Department of Energy, 2007. DOE Announces Final Rule for Loan Guarantee Program *Invites 16 Pre-Applicants to Submit Applications* for Federal Support of Innovative Clean Energy Projects Press Release <energy.gov/news/5568.htm>

Department of Energy, 2008a. DOE Announces Loan Guarantee Applications for Nuclear Power Plant Construction. Press Release, October 2, 2008. <energy.gov/news/6620.htm>.

Department of Energy, 2008b. DOE Announces Solicitations for $30.5 Billion in Loan Guarantees, Press Release, June 30, 2008 <energy.gov/news/6377.htm>.

Department of Energy, 2008c. Global Nuclear Energy Partnership. <gnep.energy.gov>.

Department of Housing and Urban Development, 2006. American Housing Survey for the United States: 1989-2005 (Washington, DC:), Table 2-24 <census.gov/hhes/www/ahs.html>.

Doheny-Farina, Stephen, 2001. *The Grid and the Village: Losing Electricity, Finding Community, Surviving Disaster.* New Haven CT: Yale University Press.

Energy Information Administration, Office of Integrated Analysis and Forecasting, 2008. Annual energy Outlook 2009. Washington, DC: Department of Energy <eia.doe.gov/oiaf/aeo>.

Energy Information Administration, 2008a. Annual Electric Generator Report, Form EIA-860. Washington, DC: Department of Energy <eia.doe.gov/cneaf/electricity/page/eia860.html>

Energy Information Administration, 2008b. *Annual Energy Review 2007*, Report No. DOE/EIA-0384, Washington, DC: Department of Energy <eia.doe.gov/emeu/aer/>.

Environmental Protection Agency, 2008. *Inventory of U.S. Greenhouse Gas Emissions and Sinks: 1990-2006.* <epa.gov/climatechange/emissions/usinventoryreport.html>.

Freese, Barbara, Steve Clemmer, and Alan Nogee, 2006. *Coal Power in a Warming World: A Sensible Transition to Cleaner Energy Options.* Cambridge MA: Union of Concerned Scientists <ucsusa.org/assets/documents/clean_energy/Coal-power-in-a-warming-world.pdf>.

George, Susan, 2003. *The Lugano Report: On Preserving Capitalism in the Twenty-First Century.* London: Pluto Press.

Geothermal Technologies Program, 2008. About Geothermal Electricity. National Renewable Energy Laboratory, Department of Energy <nrel.gov/geothermal>.

Gilroy, John Martin and Joe Bowersox, Eds., 2002. *The Moral Austerity of Environmental Decision Making: Sustainability, Democracy, and Normative Argument in Policy and Law.* Durham NC: Duke University Press.

Goodell, Jeff, 2007. *Big Coal: Dirty Secret Behind America's Energy Future.* Boston: Houghton-Mifflin.

Goodstein, David, 2005. *Out of Gas: The End of the Age of Oil.* New York, W.W. Norton.

Haller, Stephen F., 2002. *Apocalypse Soon? Wagering on Warnings of Global Catastrophe.* Montreal: McGill-Queenn's University Press.

Hansen, James E. 2007. How We Can Avert Dangerous Climate Change. Testimony to the Select Committee on Energy Independencenand Global Warming, United States House of Representatives on 26 April. <arxiv.org/ftp/arxiv/papers/0706/0706.3720.pdf>

Hansen, James, Makiko Sato, Pushker Kharecha, David Beerling, Robert Berner, Valerie Masson-Delmotte, Mark Pagani, Maureen Raymo, Dana L. Royer and James C. Zachos, 2008. Target Atmospheric CO2: Where Should Humanity Aim? *The Open Atmospheric Science Journal* 2:217-231.

Hermann, Laura, Ed., 2007. *Energy: Old Challenges, New Opportunities.* Washington, DC: Energy and Environment Program, The Aspen Institute <aspeninstituteorg>.

Hill, Jason, Erik Nelson, David Tilman, Stephen Polasky, and Douglas Tiffany, 2006. Environmental, economic, and energetic costs and benefits of biodiesel and ethanol biofuels. *PNAS* 103:11206–11210.

Holdren, John P., 2006. The Energy Innovation Imperative: Addressing Oil Dependence, Climate Change, and Other 21st Century Energy Challenges. *Innovations* 1(2):3-23 <mitpress.mit.edu/innovations>.

Interacademy Council, 2007. *Lighting the Way: Towards a Sustainable Energy Future.* <http://www.interacademycouncil.net/?id=12161>

Internal Revenue Service, 2006. Highlights of the Energy Policy Act for 2005 for Individuals <irs.gov/newsroom/article/0,,id=15>.

International Atomic Energy Agency, 2007. *Uranium 2007: Resources, Production and Demand. Organisation for Economic Co-operation and Develmopment (OECD).* London: OECD Publishing.

International Energy Agency, 2007. *Energy Technology Essentials: Nuclear Power* <http://www.iea.org/textbase/techno/essentials4.pdf>

International Energy Agency, 2008. *Energy Technology Perspectives 2008: Scenarios and Strategies to 2050.* <iea.org/textbase/npsum// ETP2008SUM.pdf>.

Intergovernmental Panel on Climate Change, 2008. Climate Change 2007. *IPCC Fourth Assessment Report.* <ipcc.ch>.

Jacobson, Mark Z., 2009. Review of Solutions to Global Warming, Air Pollution, and Energy Security. *Energy and Environmental Sciences,* Advance Article DOI: 10.1039/b809990.

Koplow, Doug, 2007. Biofuels – At What Cost? Government support for ethanol and biodiesel in the United States: 2007 Update. Global Subsidies Initiative of the International Institute for Sustainable Development, Geneva, Switzerland.

Laidi, Zaki, translated by Chris Turner, 2007. *The Great Disruption.* Cambridge: Polity Press.

Lenton, Timothy M., Hermann Held,Elmar Kriegler, Jim W. Hall, Wolfgang Lucht, Stefan Rahmstorf, and Hans Joachim Schellnhubert, 2008. Tipping elements in the Earth's climate system, *PNAS* 105 (6) 1786-93.

Lewens, Tim, Ed., 2007. *Risk: Philosophical Perspectives*, New York: Routledge.

Lovelock, James, 2006. *The Revenge of Gaia: Earth's Climate Crisis and the Fate of Humanity,* New York NY: Basic Books.

Lumb, Judy, 2008. Riding the Rails to an Energy-Efficient Transportation Future. *Quaker Eco-Bulletin* 8 (1):1-4.

Makhijani, Arjun, 2008. *Carbon-Free and Nuclear-Free: A Roadmap for U.S. Energy Policy.* Takoma Park MD: Institute for Energy and Environmental Research <ieer.org/carbonfree/CarbonFreeNuclearFree.pdf>.

Massachusetts Institute of Technology, 2003. *The Future of Nuclear Power* <web.mit.edu/nuclearpower>

Mooney, Chris, 2006. *The Republican War on Science.* New York NY: Basic Books.

Mudd, Gavin M., and Mark Diesendorf, 2008. Sustainability of Uranium Mining and Milling:Toward Quantifying Resources and Eco-Efficiency. *Environ. Sci. Technol.* 42 (7), 2624-2630 • DOI: 10.1021/es702249v

Nadeau, Robert , 2006. *The Environmental Endgame: Mainstream Economics, Ecological Disaster, and Human Survival.* New Brunswick NJ: Rutgers University Press.

National Commission on Energy Policy, 2004. *Ending the Energy Stalemate: A Bipartisan Strategy to Meet America's Energy Challenges.* <energycommission.org>.

Northcott, Michael, 2007. *A Moral Climate. The Ethics of Global Warming.* Maryknoll NY: Orbis Books.

Nuclear Regulatory Commission, 2008. *Expected New Nuclear Power Plant Applications.* <nrc.gov/reactors/new-licensing/new-licensing-files/ expected-new-rx-applications.pdf>

Oak Ridge National Laboratory, 2001. *New Concepts for the Transmission Grid.* Department of Energy Workshop on Analysis and Concepts to Address Electric Infrastructure Needs <ornl.gov/sci/htsc/documents/pdf/roadmap080301/dale.pdf>.

Patterson, Walt, 2007. *Keeping the Lights On: Towards Sustainable Electricity.* London: Earthscan <www.waltpatterson.org>.

Ronlund, Lisbeth, David Lochbaum, and Edwin Lyman. 2007. *Nuclear Power in a Warming World: Assessing the Risks, Addressing the Challenges.* Cambridge MA: Union of Concerned Scientists <ucsusa.org/assets/documents/nuclear_power/nuclear-power-in-a-warming-world.pdf>.

Scheer, Herman, 2007. *Energy Autonomy: The Economic, Social, and Technological Case for Renewable Energy.* London: Earthscan.

Simms, Andrew, 2005. *Ecological Debt: The Health of the Planet and the Wealth of Nations.* London: Pluto Press.

Spixa, Claudia, Sven Schmiedela, Peter Kaatscha, Renate Schulze-Ratha, Maria Blettnerb, 2008. Case–control study on childhood cancer in the vicinity of nuclear power plants in Germany 1980–2003. *European Journal of Cancer* 44: 275-284.

Stovall, John P., John P. Bowles, Clifford C. Diemond, Russell A. Eaton, Paul A. Gnadt, Stanley V. Heyer, Michael A. Lebow, Willis F. Long, James C. McIver, Eugene C. Starr, Robert L. Sullivan, Robert H. Lasseter and Reigh A. Walling, 1987. Comparison of Costs and Benefits for DC and AC Transmission <ornl.gov/info/reports/1987/3445601535443.pdf>.

Swan, Christopher, 2007. *Electric Water: The Emerging Revolution in Water and Energy.* Gabriola Island BC: New Society Publishers.

Whiteside, Kerry H., 2006. *Precautionary Politics: Principle and Practice in Confronting Environmental Risk.* Cambridge MA: MIT Press.

World Nuclear Association, 2005. Can Uranium Supplies Sustain the Gobal Nuclear Renaissance? <http://www.world-nuclear.org/reference/position_statements/uranium.html>.

Zweibel, Ken James Mason, and Vasalis Fthenakis, 2008. A Solar Grand Plan. *Scientific American* 298 (1): 64-73.

Quaker Institute for the Future

Mission

The mission of QIF is to advance a global future of inclusion, social justice, and ecological integrity through participatory research and discernment.

The focus of the Institute's concerns include:
- Economic behavior that increasingly undermines the ecological processes on which life depends.
- The development of technologies and capabilities that hold us responsible for the future of humanity and the Earth.
- Structural violence and lethal conflict arising from the pressures of change, increasing inequity, concentrations of power and wealth, declining natural capital, and increasing militarism.
- The increasing separation of people into areas of poverty and wealth, and into social domains of aggrandizement and deprivation.
- The philosophy of individualism and its socially corrosive promotion as the principal means for the achievement of the common good.
- The complexity of global interdependence and its demands on governance systems and citizen's responsibilities.
- The convergence of ecological and economic breakdown into societal disintegration.

The quality of the human future, and to a large extent, the future of Earth as a commonwealth of life, depends on the advance and application of social and ecological intelligence. The Religious Society of Friends (Quakers) has long carried an active concern for the application of scholarly and scientific knowledge to the great work of human betterment. Within this heritage, QIF seeks to realize the evolutionary potential of Quaker testimonies and values.

<www.QuakerInstitute.org>

www.ingramcontent.com/pod-product-compliance
Lightning Source LLC
Chambersburg PA
CBHW031423040426
42444CB00005B/690